Start with Science

50 science activities

Malcolm Dixon

Edward Arnold

First published in Great Britain 1985 by
Edward Arnold (Publishers) Ltd
41 Bedford Square, London WC1B 3DQ

Edward Arnold (Australia) Pty Ltd
80 Waverley Road, Caulfield East
Victoria 3145, Australia

British Cataloguing in Publication Data

Dixon, Malcolm, *1946*–
Start with science: 50 science activities.
1. Science – Experiments
I. Title
507'.24 Q161.2

ISBN 0 7131 7363 7

Drawings by Margaret Theakston

Text set in Compugraphic Univers
by CK Typesetters, 26 Mulgrave Road, Sutton, Surrey
Printed and bound: Great Britain

General Aim

This pack contains 50 sheets of science activities, divided into themes, to assist the non-specialist teacher who wishes to include some practical science work in his/her classroom. Some teachers may wish to use the pack as 'dip into' material from which they can select sheets to develop specific science projects, or for the scientific section of more general topic work.

Using the sheets, which cover a wide range of subjects, the children will be introduced to the processes of scientific thinking such as observing, experimenting, looking for patterns and communicating. Other skills, particularly in the areas of mathematics and language, will also be developed.

Class Organisation

It is intended that the sheets be used in a flexible manner – by individual children, small groups, or by a whole class working on the same topic. Thorough preparation before the lesson starts is essential. Each sheet lists the materials needed for the activity: enlist the help of the children and their parents in obtaining these materials where necessary and ensure that the children have adequate working-space for their activities. As the sheets are completed, the children can build up a 'file' of their science activities. A record sheet is included so that each child can record the activities he/she has completed. At first, start with a small number of children carrying out science activities. As your, and their, confidence and experience grows, you can allow an increasing number of children to experiment at the same time.

M.D.

Contents

Materials needed

Most materials can easily be found around school or brought from home. Those marked with a single asterisk* could be purchased, found in the school science stock or borrowed from a teachers' centre or local secondary school. For those marked with a dagger† look in the school grounds, gardens or local parks.

Light
coloured paper
kaleidoscope*
paint
scissors
small mirrors*
thin card
thin string

Air
balsawood* or polystyrene
dowel rod*
drinking-straws
heavy metal nut
large balloon
measuring-tape
nylon line
paper
paperclips
plastic tray
ruler
scissors
sticky tape
thin card
thin polythene
thin string
tray of water

Water
bowl (or sink) of water
chalk
coffee
curry-powder
plastic bottle
plasticine
salt
scissors
small objects such as:
- cork
- ice-cube
- piece of polystyrene
- rubber ball
- small brick
- table-tennis ball
- wood

soap
sugar
washing-powder
washing-up liquid
yoghurt pots

Birds
binoculars (if possible)*
feathers (down and flight)†
guide book to birds
magnifying glass*

Weather
dowel rod*
graph paper
ice-cubes
large cardboard box
newspaper weather forecasts
plastic bottle
scissors
sticky tape
5 test-tubes*
thermometers*
water

Sound
bowl of water
drinking-straws
ruler
scissors
8 similar glass bottles
spoon
string
table-tennis ball
thread
tin cans
tuning-forks*

Electricity
batteries (4.5 volt)*
bulb-holders*
bulbs (3.5 volt)*
drawing-pins
paperclips
switches*
wire (insulated, but ends bared)*
wooden block

Magnetism
battery (6 volt)*
large iron nail
magnets (various types)*
paperclips
plastic bowl
scissors
short stick
switch*
thin coloured card
thin string
wire (insulated)*

Materials
clock or watch
corrugated cardboard
cotton wool
ice-cubes
newspapers
plastic bucket
polystyrene chips
sand
scissors
string
strip of wood
threads (various types)
washing-up-liquid bottle
wet cloth

Small animals
blotting paper
card
cotton wool
earthworms†
plastic tray
magnifying glass*
paper
plastic box
ruler
scissors
woodlice†

Plants
buttercups†
card
cotton wool
eggshells
egg-trays
garden flowers†
graph paper
grass seeds†
jars or vases of water
leaves†
newspapers
onion
paints and brushes
ruler
tree-trunk section†
vase with a narrow neck

Ourselves
clock or watch
drawing-pin
drinking-straw
ink-pad
magnifying glass*
ruler

Start with Science

Record Sheet

I have tried these activities: (mark with a ✓)

Name *Class* *Date*

Light 1 Reflections 2 Using mirrors 3 Make a periscope 4 Make a 'spinner'	
Air 5 Make a model parachute 6 A 'jet-powered' balloon 7 Test a sailing boat 8 Make a model hovercraft 9 Make paper helicopters	
Water 10 Floating and sinking 11 Making boat shapes 12 Dissolving in water 13 Giant bubbles	
Birds 14 Looking at birds 15 Feathers 16 Birds flying	
Weather 17 Using a thermometer 18 The week's weather 19 How strong is the wind? 20 How much rain?	
Sound 21 Vibrations 22 Tuning-forks 23 Make a telephone 24 Play a tune 25 Sounds from a straw pipe	

Electricity 26 Make an electric circuit 27 Use a switch in a circuit 28 More circuits 29 Morse-code messages	
Magnetism 30 Using magnets 31 A magnetic fishing game 32 Make an electromagnet	
Materials 33 Strong threads 34 Make a 'spin-drier' 35 Fun with ice-cubes	
Small animals 36 Looking at earthworms 37 It's moving! 38 Woodlice (1) 39 Woodlice (2)	
Plants 40 Collecting leaves (1) 41 Looking at leaves (2) 42 Tree shapes 43 Annual rings 44 Growing grass seeds 45 Flowers 46 Grow an onion in water	
Ourselves 47 Eyes 48 Looking at fingerprints 49 Measure your pulse 50 Test your reactions	

An Edward Arnold Master

Start with Science

Light:

Reflections

Name *Class* *Date*

■ **You need:** ☐ small mirror

Look in a mirror.
Can you see your reflection?
Touch your right ear.
Which ear is your reflection touching?

Write your name on a piece of paper.
Hold your writing in front of the mirror.
Look at the reflections of the letters in the mirror.

What do you see?

Why do these people use mirrors? Write a sentence about each one.

Car drivers

Hairdressers

Dentists

Look for reflections in shiny spoons, in pools of water and in sides of polished cars.

An Edward Arnold Master

2

Start with Science

Light:

Using mirrors

Name *Class* *Date*

■ **You need:** ☐ 3 small mirrors ☐ coloured paper ☐ kaleidoscope

Find three small mirrors.
Stand them on your desk.
Make a triangle shape.

Cut some small pieces of coloured paper.
Drop these pieces inside the triangle shape.
Look in the mirrors.
What can you see?

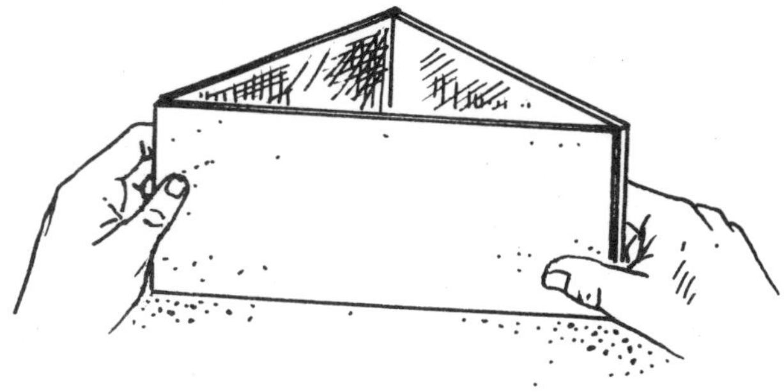

Find a kaleidoscope.
Look inside. What do you notice?

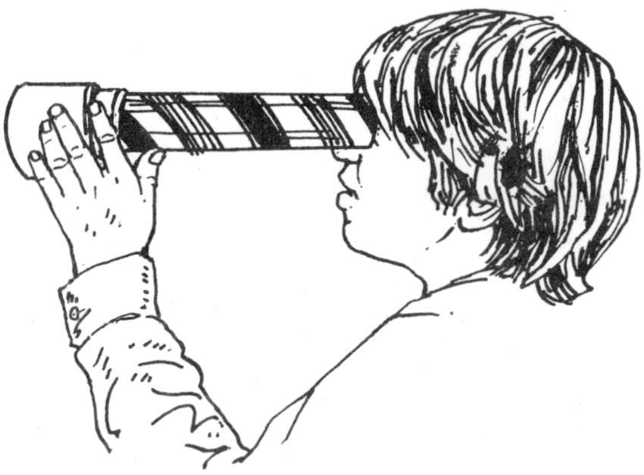

How do you think a kaleidoscope works?

An Edward Arnold Master

Start with Science

3 Light:

Make a periscope

Name *Class* *Date*

■ **You need:** ☐ 2 small mirrors

Sit on a chair.
Hold a mirror in front of your eyes.

Ask a friend to hold a second mirror above the mirror in your hand.

Look at your mirror.

Can you see the second mirror?
Can you see what is happening behind your back?

When two mirrors are arranged like this they make a **periscope**.
Periscopes can be used to see over crowds.
They are also used in submarines.

Use the word **periscope** in two sentences of your own.

An Edward Arnold Master

4

Start with Science

Light:

Make a 'spinner'

Name *Class* *Date*

■ **You need:** ☐ thin card ☐ paint ☐ scissors ☐ thin string

Draw a circle on a piece of thin card.
Cut out the circle shape.
Divide each side into half.
Paint each half a different colour.

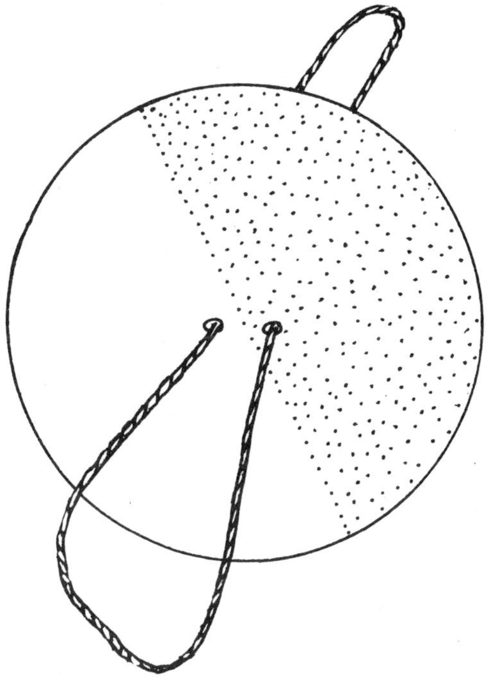

Make two small holes in the centre of the card.
Thread some thin string through these holes.
Use the string to spin the card.
What colour is the spinning card?

____

Make more spinners. Paint them different colours.
Write down the colours they are to start with.
Write down the colours you observe when they are spinning.

Colours to start with	Colours when spinning

An Edward Arnold Master

5

Start with Science

Air:

Make a model parachute

Name *Class* *Date*

■ **You need:** ☐ thin polythene ☐ thin string ☐ heavy metal nut ☐ scissors

Cut a square piece of thin polythene.
It should measure about 30 centimetres along each edge.

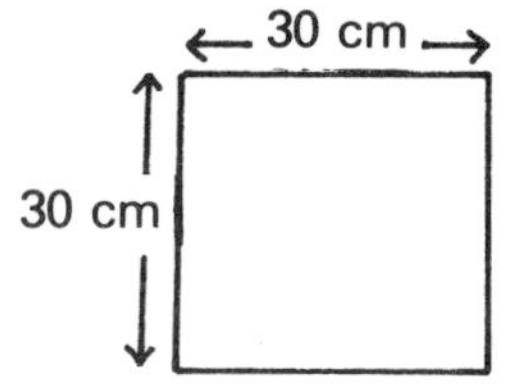

Tie lengths of thin string to each corner.
Fasten a heavy metal nut to the ends of the string.

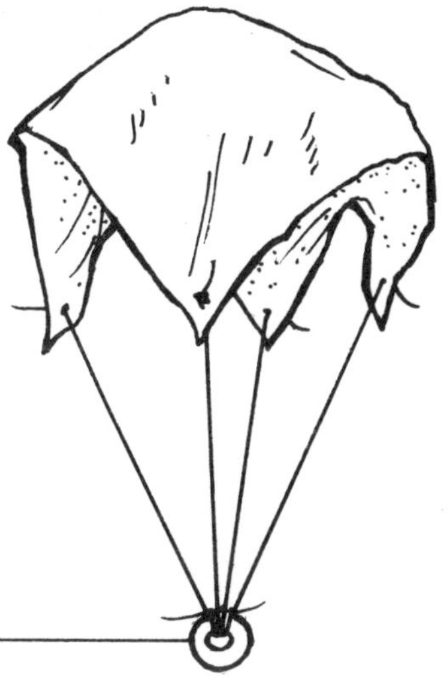

Go into the school playground.
Throw your parachute into the air.
Watch how it travels down to the ground.
Why does it travel slowly?

The parachute travels slowly because ______________________

____

____

Cut a small hole in your model parachute.
Test it again.
Does the hole make any difference?
Why do you think this is so?

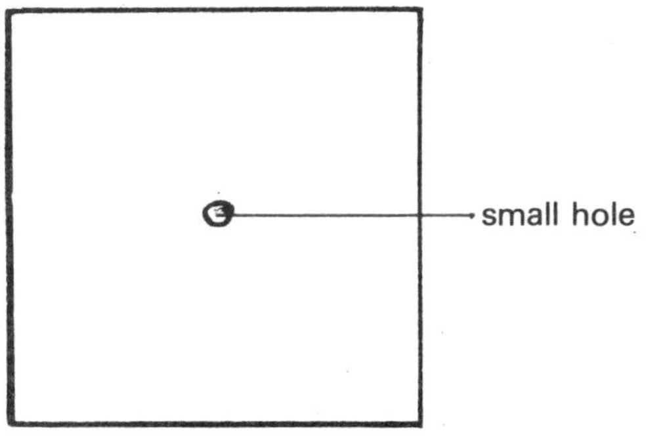

____

____

____

____

An Edward Arnold Master

6

Start with Science

Air:

A 'jet-powered' balloon

Name *Class* *Date*

■ **You need:** ☐ large balloon ☐ sticky tape ☐ drinking-straw ☐ nylon line ☐ measuring-tape

Find a large balloon.
Use sticky tape to fasten a short drinking-straw to the balloon.
Pass a length of nylon line through the straw.

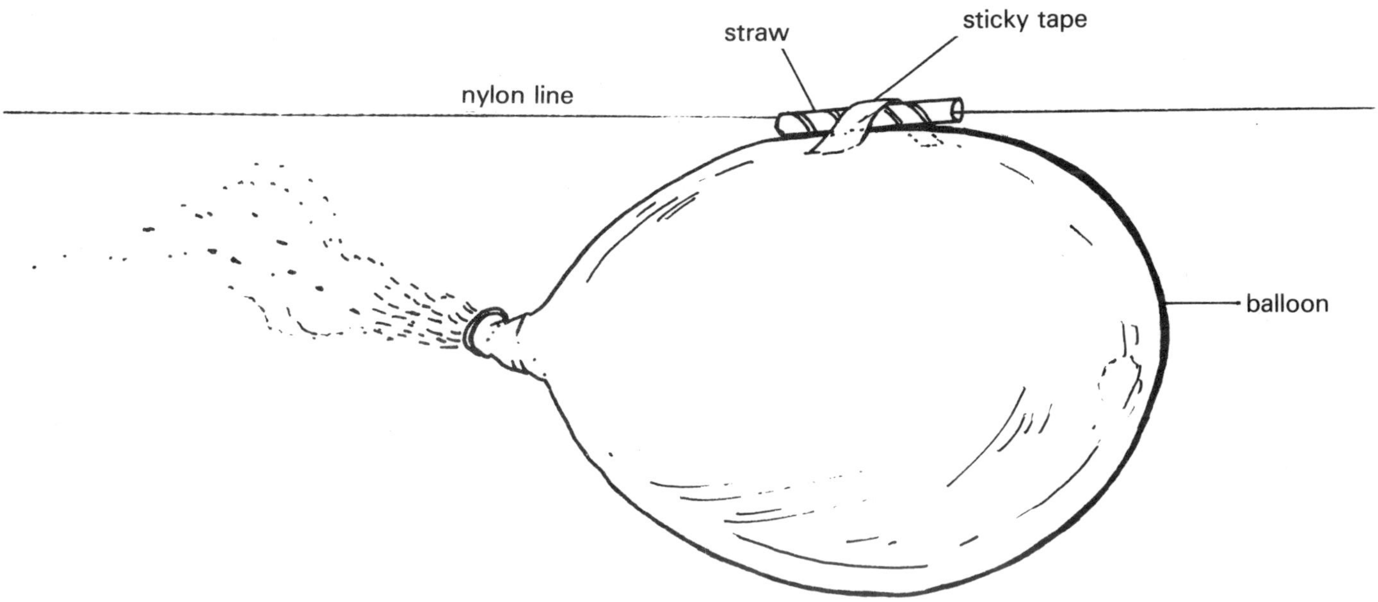

Blow up the balloon. Hold the end.
Let your balloon go. Measure how far it travels.

The balloon moved ______ cm.

Try again. How far does it travel this time?

It travelled ______ cm.

Now complete these sentences:

A __ __ is forced out of the b __ __ __ __ __ __ . As it leaves it pushes the balloon in the

opp __ __ __ __ __ direction.

Collect some pictures of rockets and jet-powered aeroplanes.

An Edward Arnold Master

7	*Start with Science*

Air:
Test a sailing boat

Name *Class* *Date*

■ **You need:** ☐ balsawood or polystyrene ☐ dowel rod ☐ thin card ☐ tray of water ☐ drinking-straw ☐ scissors

Make a boat shape using a small block of balsawood or polystyrene.
Push a length of dowel rod into the centre of the boat.
Make some different-shaped sails using thin card.
Fit one sail onto the dowel mast.

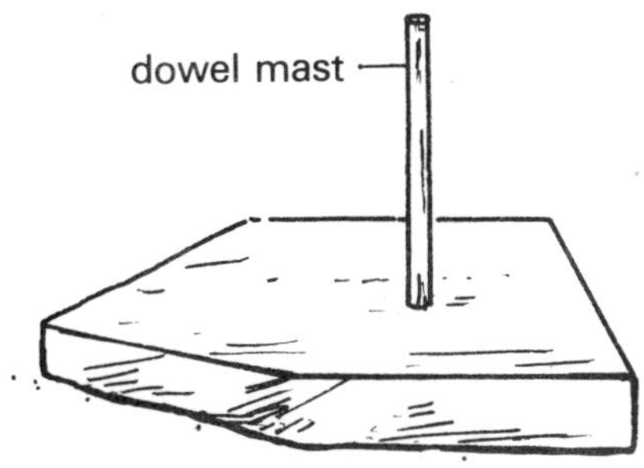

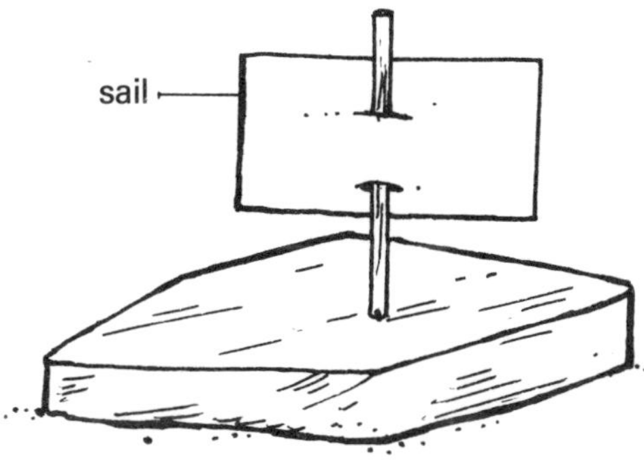

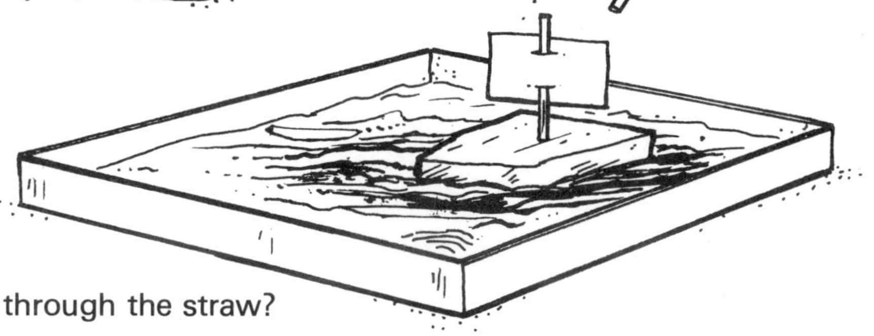

Float your boat on a tray of water.
Blow through a straw at the sail.
How far does the boat move?
Fit the other sails in turn to your boat.
How far does it move on one blow of air through the straw?

Shape of sail	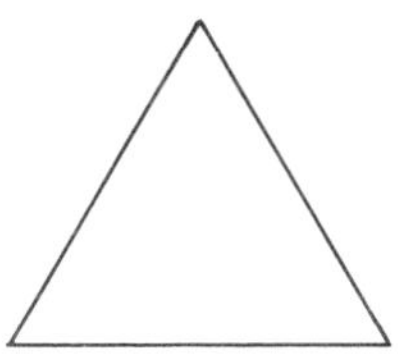	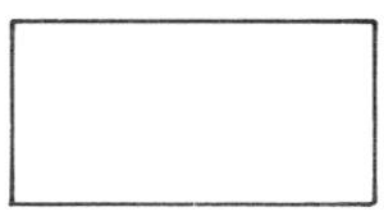			
Distance moved	cm	cm	cm	cm	cm

Which is the best sail? ______

Can you suggest why? ______

______

An Edward Arnold Master

8	*Start with Science* **Air:** **Make a model hovercraft**

Name *Class* *Date*

■ **You need:** ☐ plastic tray ☐ scissors ☐ thin card

Find an empty plastic tray.
Make a small hole in the tray.
Make a small funnel using thin card.
Fit the funnel through the small hole in the tray.

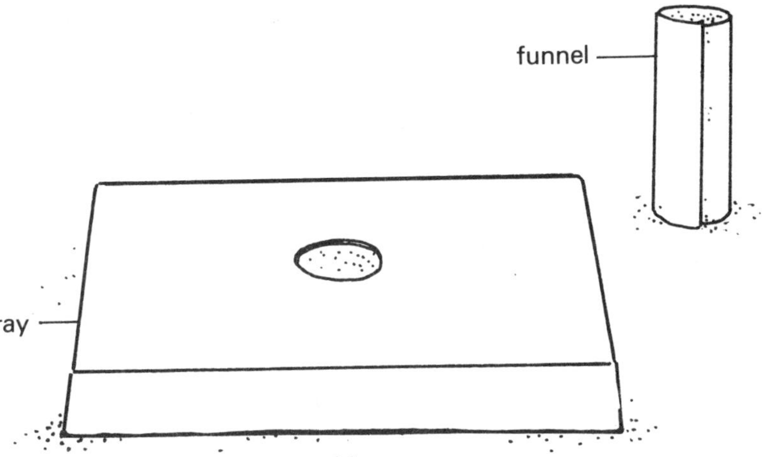

Place your model on a table top.
Blow through the funnel.
Does your model glide over the surface of the table?
If not, try using a lighter tray.

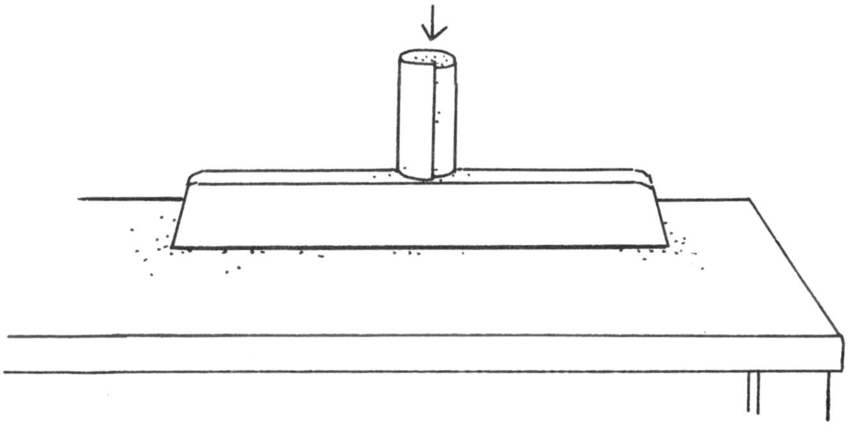

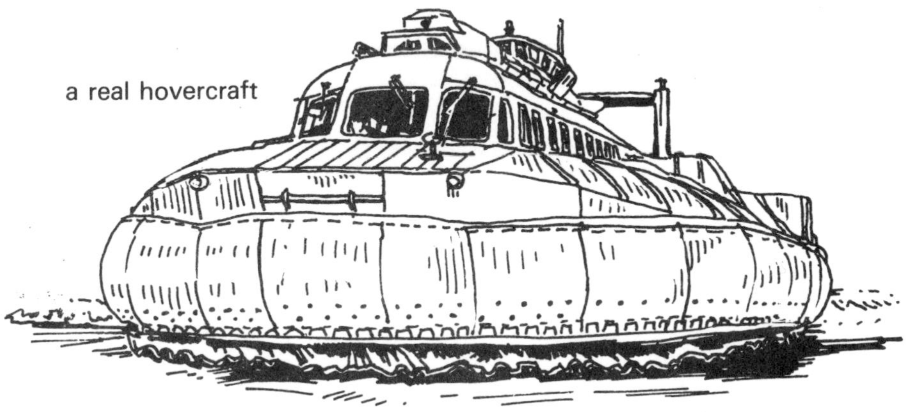

Complete these sentences:

Real h _ _ _ _ _ _ _ _ _ also glide on a cushion of _ _ r.

Powerful motors are used to make this air cushion.

Try to find out more about hovercraft.

An Edward Arnold Master

Start with Science

9 Air:

Make paper helicopters

Name Class Date

■ **You need:** ☐ paper ☐ ruler ☐ scissors ☐ paperclips

Cut six pieces of paper so that each measures 12 cm by 4 cm. Measure 2 cm across the top of the paper and mark this point. Cut down from this mark and fold back the wings. Do this for all six pieces of paper. Make each cut a different length. Fix a paperclip to each helicopter.

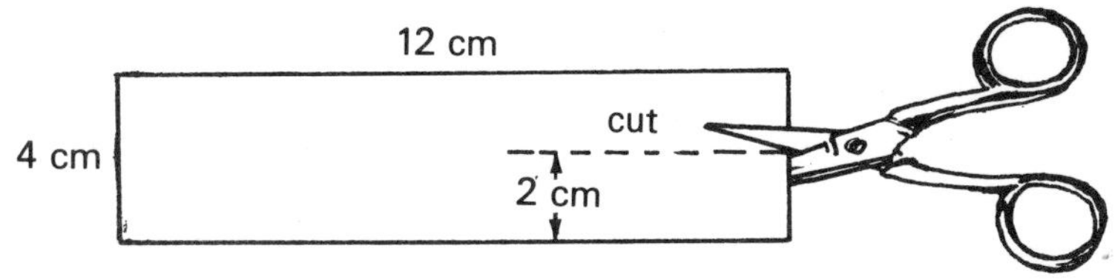

Each helicopter should look like this.

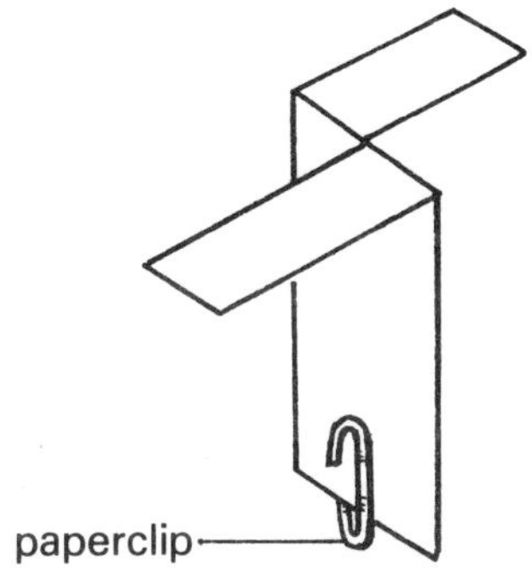

Which helicopter falls the quickest? ______

______

Which helicopter takes the most time to fall? ______

______

What happens when the wings are shortened? ______

______

What happens when more paperclips are attached? ______

______

An Edward Arnold Master

10

Start with Science

Water:

Floating and sinking

Name *Class* *Date*

■ **You need:** ☐ bowl (or sink) of water and many small objects such as:
☐ piece of polystyrene ☐ wood ☐ cork ☐ rubber ball ☐ small brick
☐ table-tennis ball ☐ ice-cube

Collect many objects made from different materials. Try to include a piece of polystyrene, some wood, a cork, a rubber ball, a small brick and a table-tennis ball in your collection.

Guess which objects will float in water.

Test them in a sink, or large bowl of water.

Record your results:

Float	Sink

Does an ice-cube float or sink? ______________________________

____

An Edward Arnold Master

11

Start with Science

Water:

Making boat shapes

Name *Class* *Date*

■ **You need:** ☐ plasticine ☐ bowl (or sink) of water

Take a ball of plasticine.
Do you think it will float or sink in water?

Test it to find out.
What happens?

Take the ball of plasticine and make a boat shape.

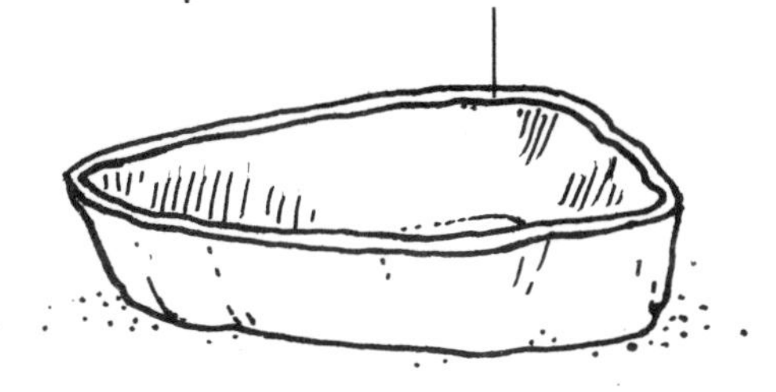

What happens when you place this shape in water?
How much 'cargo' will your plasticine boat carry?
Make other boat shapes.
Can they carry more cargo?

Describe your experiments:

What materials are used to build modern ships?

An Edward Arnold Master

12

Start with Science

Water:

Dissolving in water

Name *Class* *Date*

■ **You need:** ☐ yoghurt pots ☐ salt ☐ washing-powder ☐ soap ☐ sugar ☐ coffee ☐ curry-powder ☐ chalk ☐ water (cold)

Collect some salt, washing-powder, soap, sugar, coffee, curry-powder and chalk.
Put each substance into a clean yoghurt pot.
Put some cold water into other yoghurt pots.
Stir a small amount of salt into the water in one pot.
Does it **dissolve** in the cold water?
Try the other substances.

Complete this list:

Dissolves in cold water	Does not dissolve in cold water

An Edward Arnold Master

13

Start with Science

Water:

Giant bubbles

Name *Class* *Date*

■ **You need:** ☐ plastic bottle ☐ scissors ☐ washing-up liquid ☐ water (warm) ☐ bowl (or sink)

Find an empty plastic bottle.
Make sure it is clean.
Cut the bottle in half.
Cut some slits around the widest end.
Bend these pieces back.

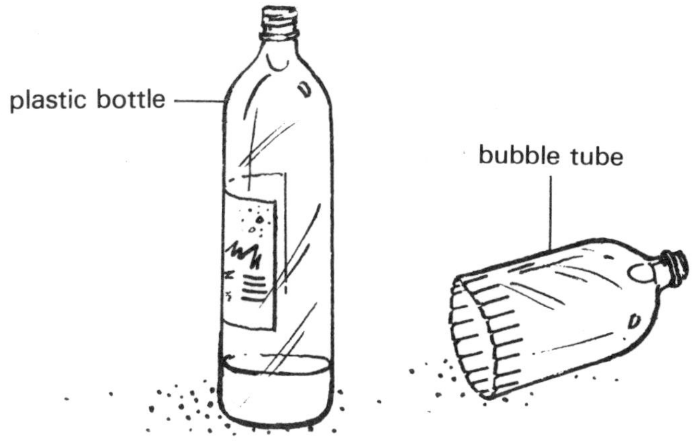

Mix a little washing-up liquid with some warm water.
Dip your plastic bubble tube into this solution.
Pull out the tube.
Blow gently through the tube to form a giant bubble.
Watch it float through the air.

Blow more giant bubbles.
What is inside these bubbles?

What shape are your bubbles?

How could you measure the **diameter** of a bubble?

An Edward Arnold Master

14

Start with Science

Birds:

Looking at birds

Name *Class* *Date*

■ **You need:** ☐ guide book to birds

Find a good spot to watch birds.
Remember to stay quiet and keep still.
Record what you notice about the birds.
Use a bird book to help you identify them.

sparrow

robin

Date ______ Place ______

How many birds can you see? ______

What are they doing? ______

What colours are they? ______

What sounds do they make? ______

Watch one bird for fifteen minutes. Record what it did in this time:

An Edward Arnold Master

Start with Science

15

Birds:

Feathers

Name Class Date

■ **You need:** ☐ feathers (down and flight) ☐ magnifying glass

Birds have different types of feathers.
The **down feathers** keep the bird warm.
The **flight feathers** help the bird to fly.

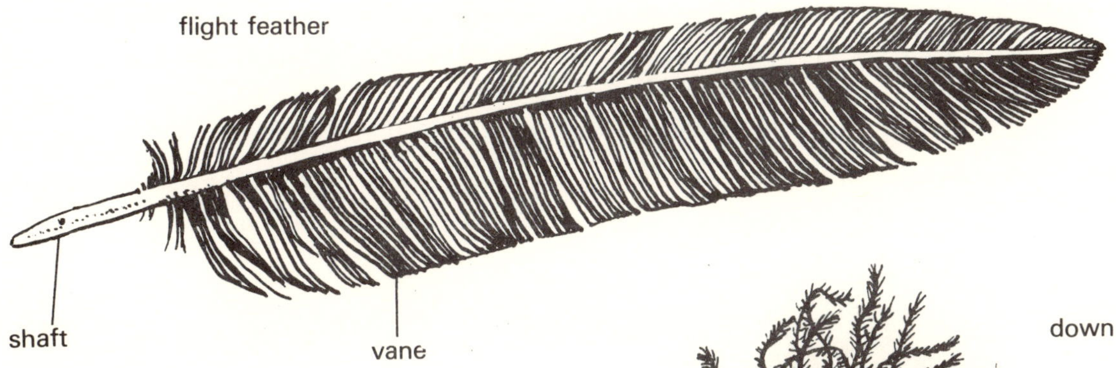

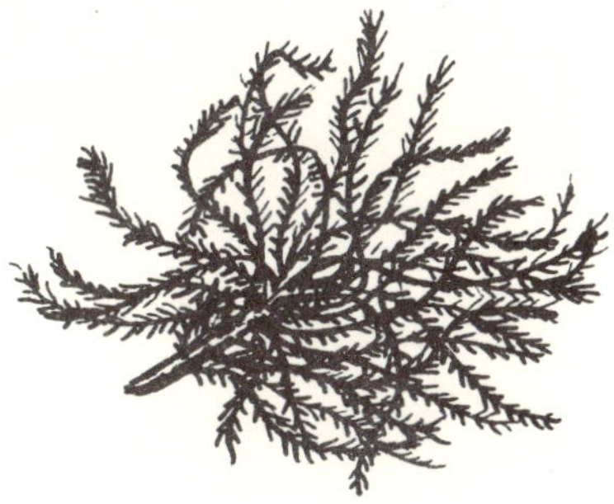

Collect some feathers.
Make a drawing of a flight feather.
Label the **shaft** and the **vane**.

Examine the vane with a magnifying glass.
What do you notice?

Watch some birds flying. Watch how the birds use their **wings** to push through the air.

An Edward Arnold Master

Start with Science

16

Birds:

Birds flying

Name Class Date

■ **You need:** ☐ binoculars (if possible)

Some birds fly vast distances every year.
Try to find out *why* they fly such distances.
How do they find their way?

Lesser black-backed gull

Watch some birds flying.
Watch how they take off.

What do you notice?

Do all birds take off the same way?

Watch the patterns the birds make in the air.
Draw some of these patterns.

Watch how they use their wings.
Do all birds **glide** through the air?

Watch how birds land.

An Edward Arnold Master

17	*Start with Science*

Start with Science

Weather:
Using a thermometer

Name	Class	Date

■ **You need:** ☐ thermometer ☐ water (hot and cold) ☐ ice-cubes

Your teacher will give you a **thermometer**.
Handle it very carefully.
It will tell you the **temperature**.

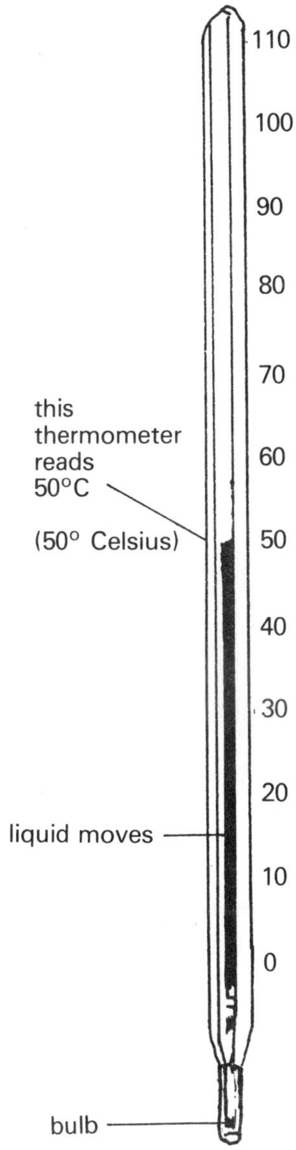

Place the **bulb** of your thermometer in some cold water.
Watch the liquid inside the thermometer move.
What is the temperature of the cold water?

Use your thermometer to measure other temperatures.

Your classroom	______ °C
Some hot water (take care)	______ °C
In the playground	______ °C
Some ice-cubes	______ °C

An Edward Arnold Master

18

Start with Science

Weather:

The week's weather

Name *Class* *Date*

■ **You need:** ☐ newspaper weather forecasts

Keep a record of the weather every day for a school week. This chart will help. Draw pictures of the weather and describe it in your own words. Look at some newspaper weather forecasts. How do they compare with your records?

	Picture	Description	Weather forecast
Monday			
Tuesday			
Wednesday			
Thursday			
Friday			

An Edward Arnold Master

19

Start with Science

Weather:

How strong is the wind?

Name *Class* *Date*

■ **You need:** ☐ large cardboard box ☐ scissors ☐ dowel rod ☐ sticky tape

Find an empty cardboard box.
Cut away both ends.
Fix a length of dowel rod
through one end.
Make a cardboard flap and
attach it to the dowel rod.
It should swing easily.

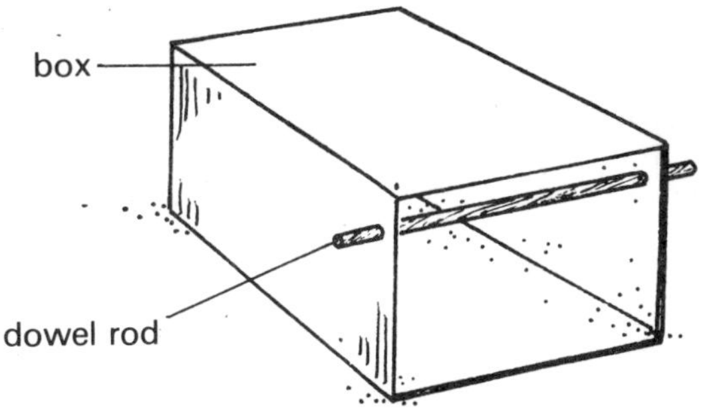

Hold your wind-speed recorder
so that the flap faces the wind.
Does the wind push the flap into the box?

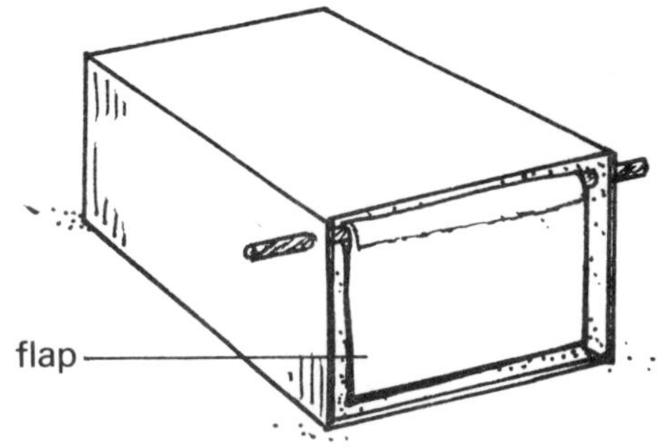

Use your recorder to measure
the wind speed on different days.

Describe what happens:

20

Start with Science

Weather:

How much rain?

Name *Class* *Date*

■ **You need:** ☐ plastic bottle ☐ scissors ☐ 5 test-tubes ☐ graph paper

Use a plastic bottle to make a rain-gauge.
The drawings show you how:

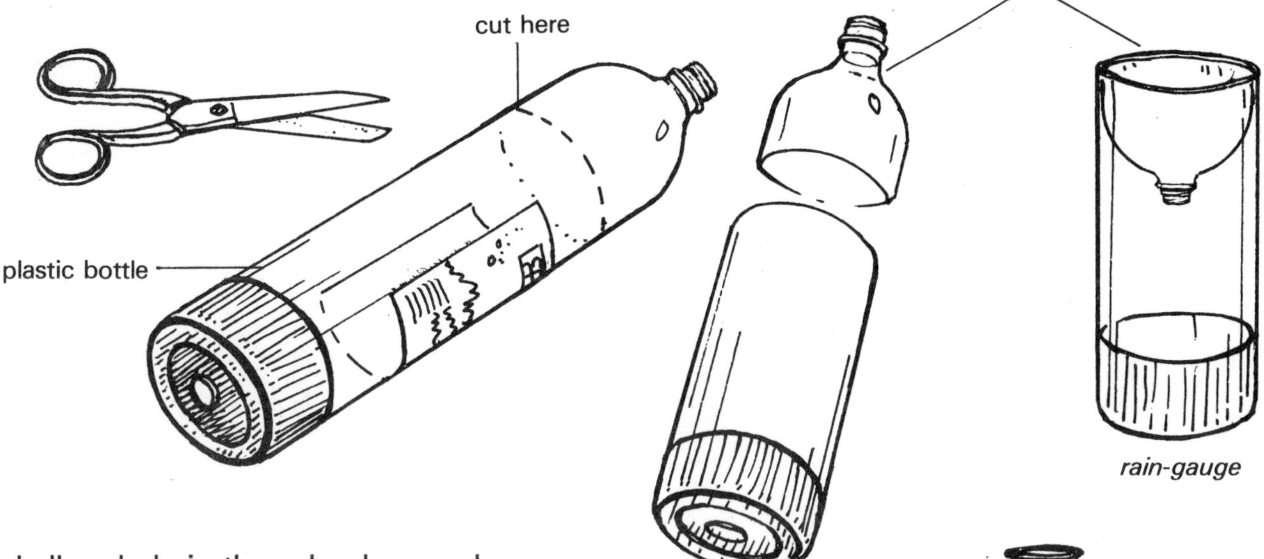

Dig a shallow hole in the school grounds.
Position your rain-gauge in this hole.
Make sure it is not under trees or near buildings.

Every day empty the rain that has collected into a test-tube.
Look at the different amounts collected over five days.
Draw a graph of your results.

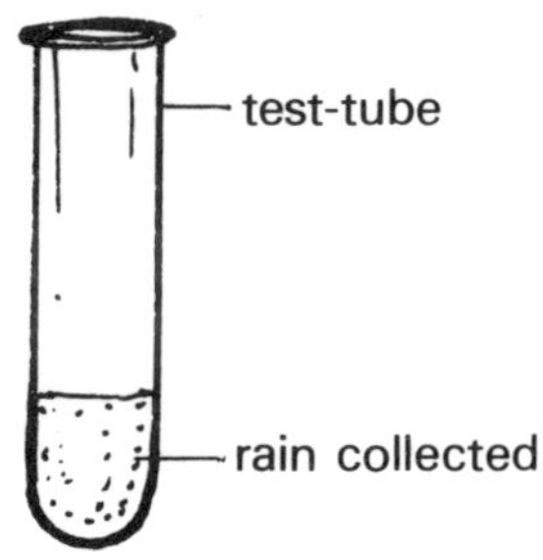

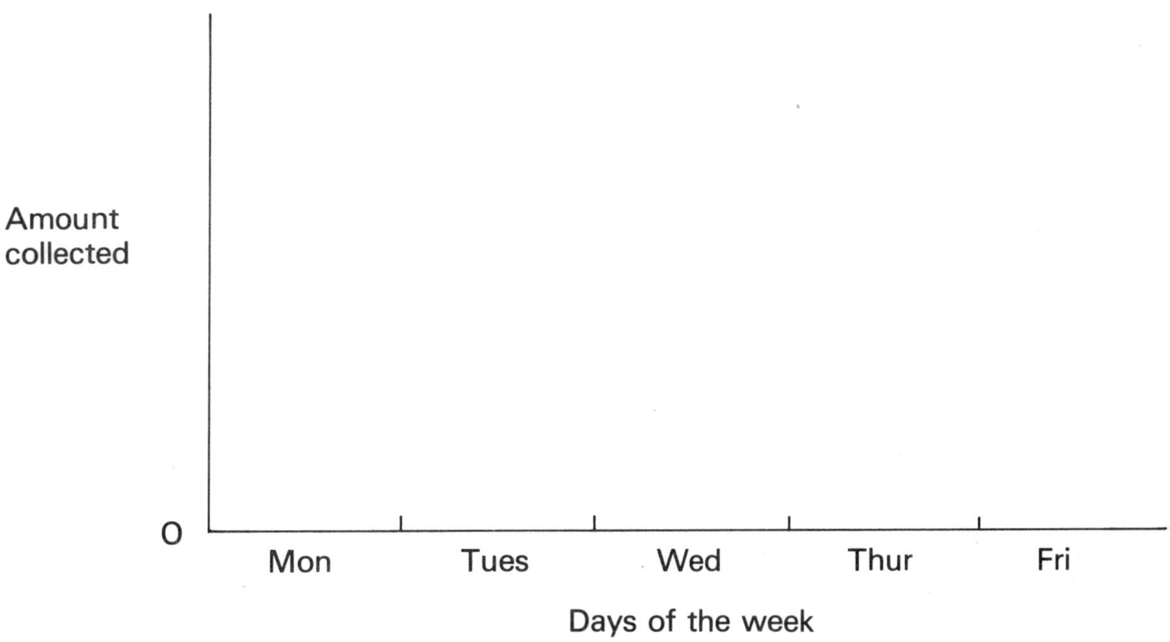

An Edward Arnold Master

21

Start with Science

Sound:

Vibrations

Name *Class* *Date*

■ **You need:** □ ruler

Place a ruler on the edge of a table.
Hold one end of the ruler.
Pull down the other end and let it go.
Watch how the ruler moves up and down.
We say the ruler is **vibrating**.
Listen to the *sound* that the vibrating ruler makes.

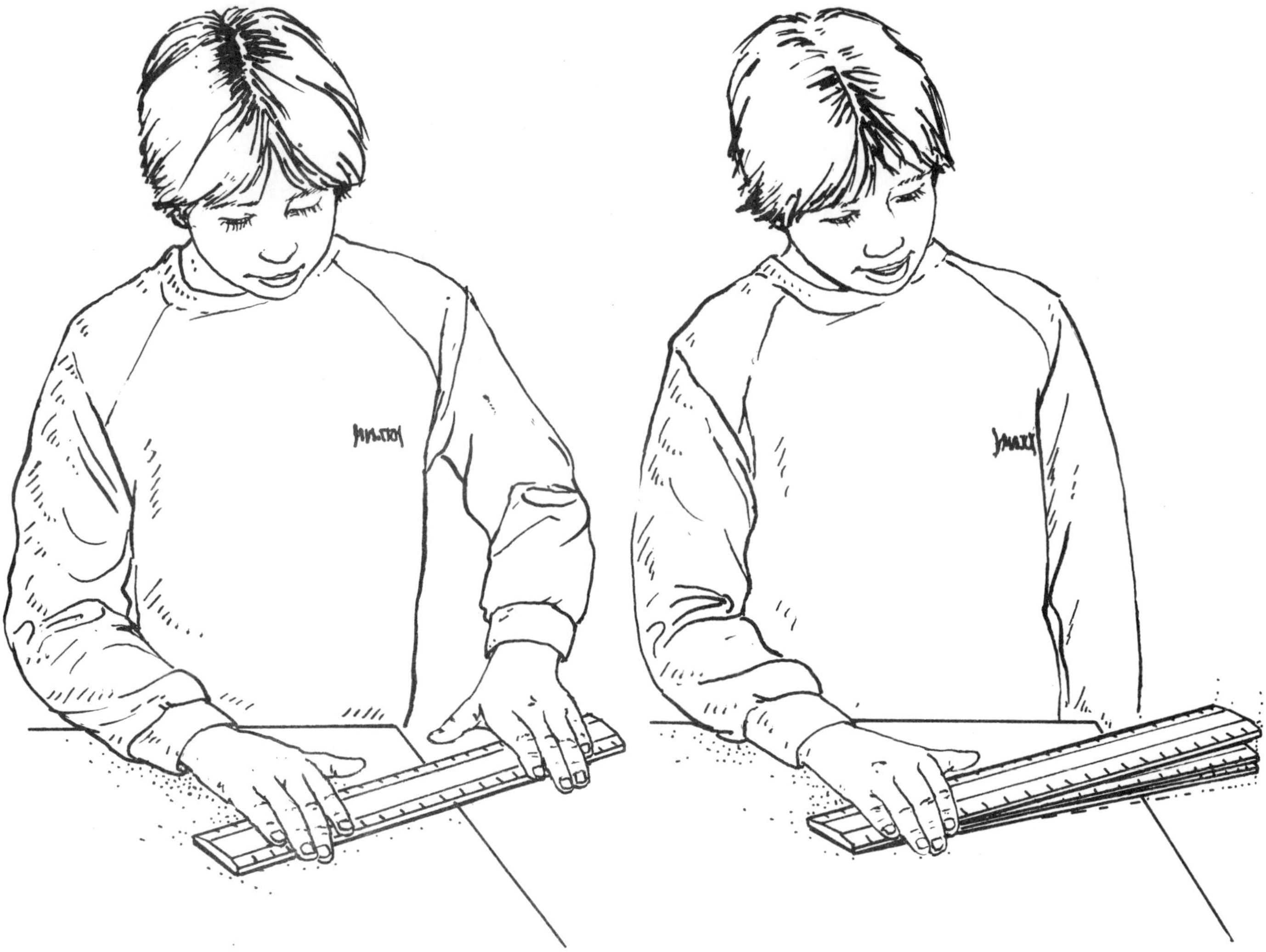

Now complete these sentences:

S _ _ _ _ _ are made by v _ _ _ _ _ _ _ _ s.

V _ _ _ _ _ _ _ _ _ move through the air to our e _ _ s.

An Edward Arnold Master

22

Start with Science

Sound:

Tuning-forks

Name *Class* *Date*

■ **You need:** ☐ tuning-forks ☐ table-tennis ball ☐ thread ☐ bowl of water

Tap a **tuning-fork** on an exercise book.
Hold the **prongs** next to your finger-nails.
Can you feel the **vibrations**?

Try these experiments. Write down your observations.

1 Hold a vibrating tuning-fork next to a table-tennis ball.

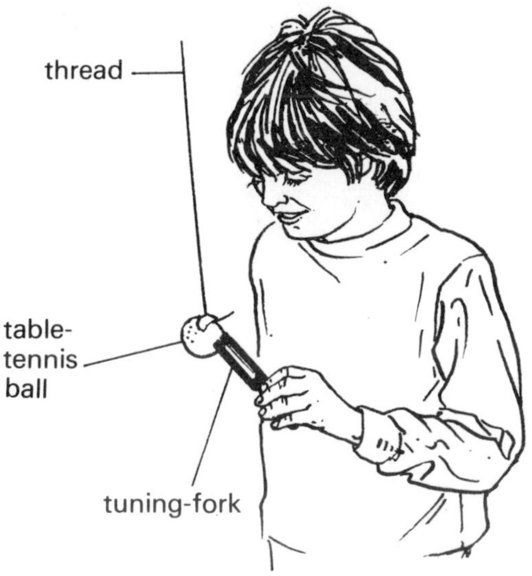

What happened?

____

____

____

____

2 Let a vibrating tuning-fork touch some water.

What happened?

____

____

____

____

An Edward Arnold Master

23	*Start with Science* **Sound:** **Make a telephone**

Name *Class* *Date*

■ **You need:** ☐ tin cans ☐ string ☐ scissors

Find two empty and clean tin cans.
Make a small hole in the bottom of each can.

Cut a length of string.
Thread one end of the string through the hole in one of the cans.
Do the same with the other can.
Tie large knots inside the cans.
Can a friend hear you speaking through the telephone?
Does it make any difference if the string is tight or slack?

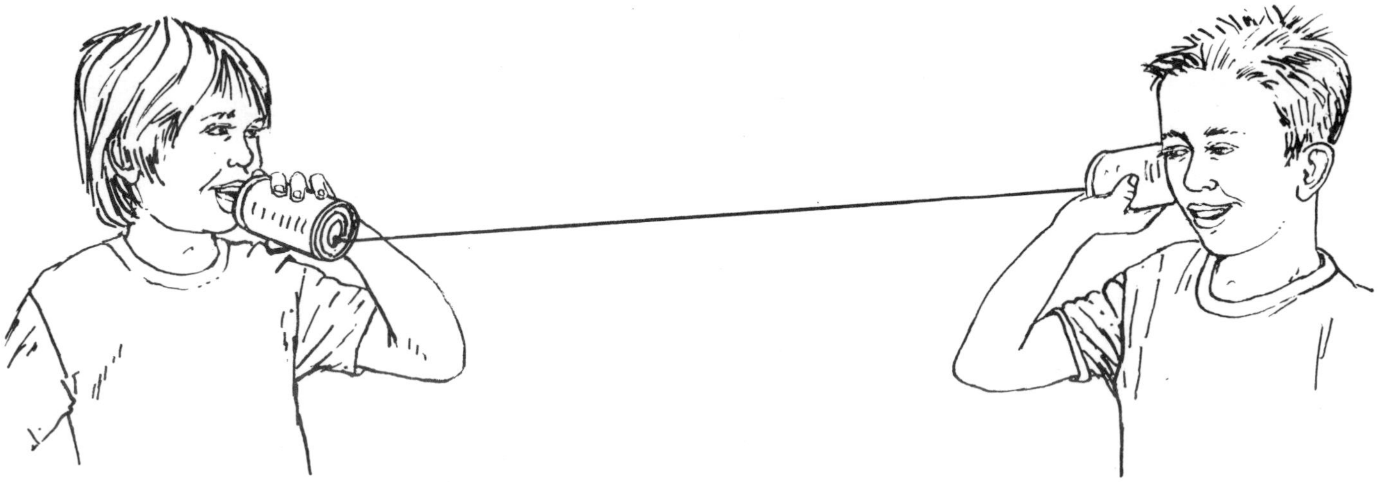

How does the sound travel from your voice to your friend's ear?

An Edward Arnold Master

Start with Science

24 Sound:

Play a tune

Name *Class* *Date*

■ **You need:** ☐ eight similar glass bottles ☐ water ☐ spoon

First you need to collect eight similar glass bottles.
Arrange the bottles in a row.
Leave the first bottle empty.
Put a small amount of water in the second.
Put slightly more water in the next bottle, and so on.
The last bottle should be filled with water.

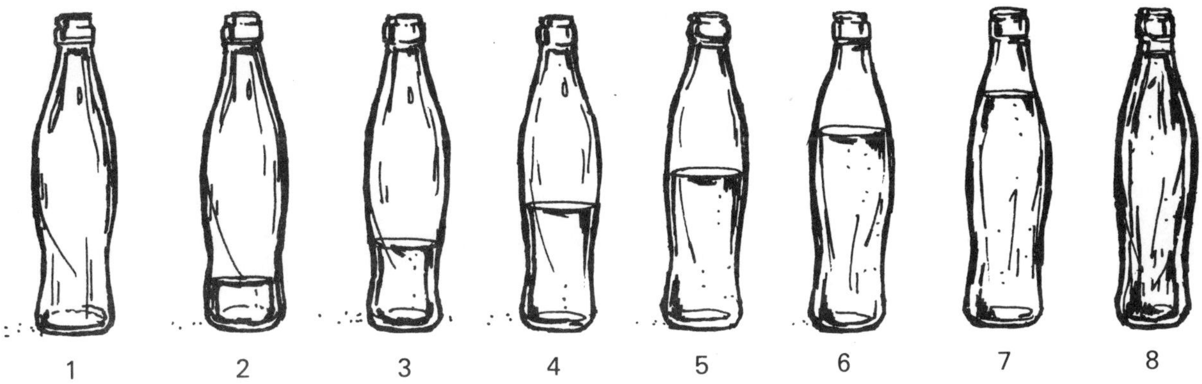

Hit the bottles gently with a spoon.
Which bottle makes the lowest note?
Which bottle makes the highest note?

Play a tune on the bottles. Record the bottle numbers below:

An Edward Arnold Master

25

Start with Science

Sound:

Sounds from a straw pipe

Name *Class* *Date*

■ **You need:** ☐ drinking-straws ☐ scissors

Take a drinking straw and flatten one end.

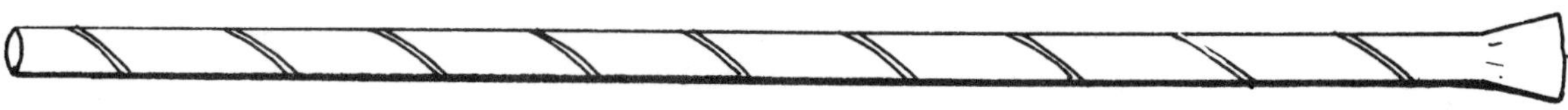

Cut the corners from the flattened end.

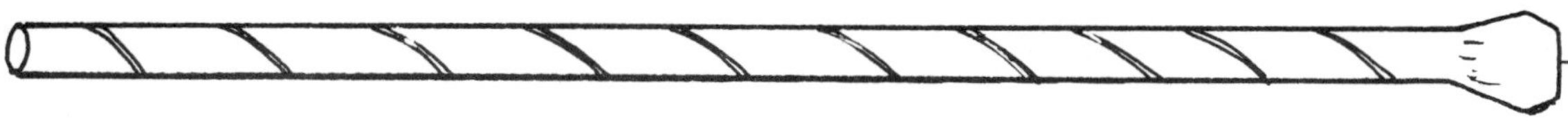

Place the straw in your mouth with the flat part above your tongue.
Blow through the straw.
Listen to the sound the straw pipe makes.

Now cut off a piece from the end of your straw pipe.

Blow again. Has the sound changed? ______

Is it lower or higher? ______

What happens to the sounds if you keep making the pipe shorter?

______

______

______

______

An Edward Arnold Master

26

Start with Science

Electricity:

Make an electric circuit

NEVER EXPERIMENT WITH MAINS ELECTRICITY.
It can kill you.

Name *Class* *Date*

■ **You need:** ☐ battery (4.5 volt) ☐ wire (insulated, but ends bared) ☐ bulb (3.5 volt) ☐ bulb-holder

Find a battery like the one in the drawing.
Look at the two metal terminals.
Connect a piece of wire to one terminal.
Connect a piece of wire to the other terminal.
Join these wires to a bulb-holder.
Screw a bulb into the bulb-holder.
What happens?

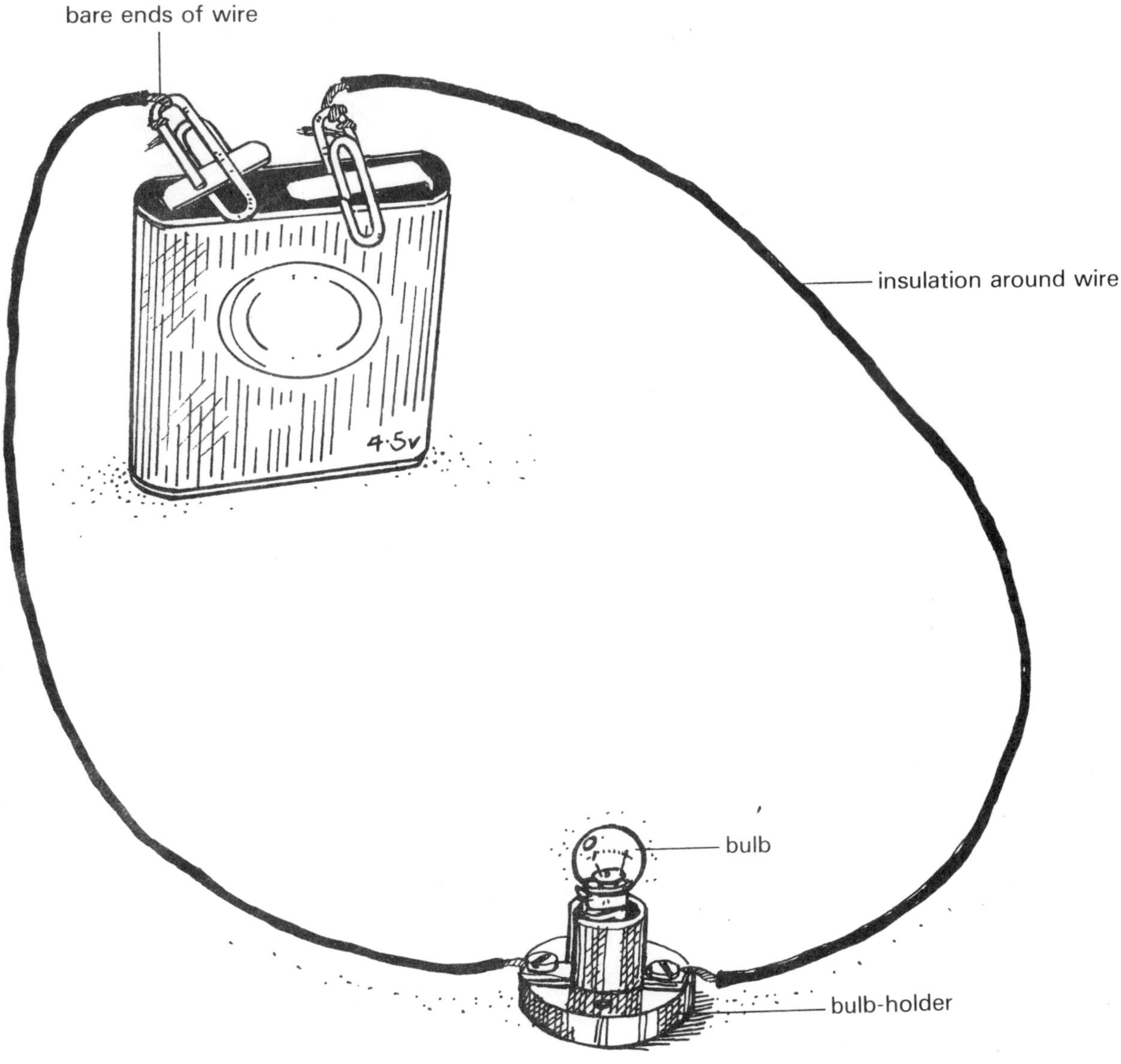

You have made an electric **circuit**.
Electricity travels from the battery, around the circuit and back to the battery.

An Edward Arnold Master

27

Start with Science

Electricity:

Use a switch in a circuit

NEVER EXPERIMENT WITH MAINS ELECTRICITY

Name *Class* *Date*

■ **You need:** ☐ wooden block ☐ drawing-pins ☐ paperclip ☐ battery (4.5 volt) ☐ bulb (3.5 volt) ☐ bulb-holder ☐ wire (insulated, but ends bared)

Find a small piece of wood, two drawing-pins and a paperclip.
Use these to make a **switch**.

Connect your switch to a battery and bulb.

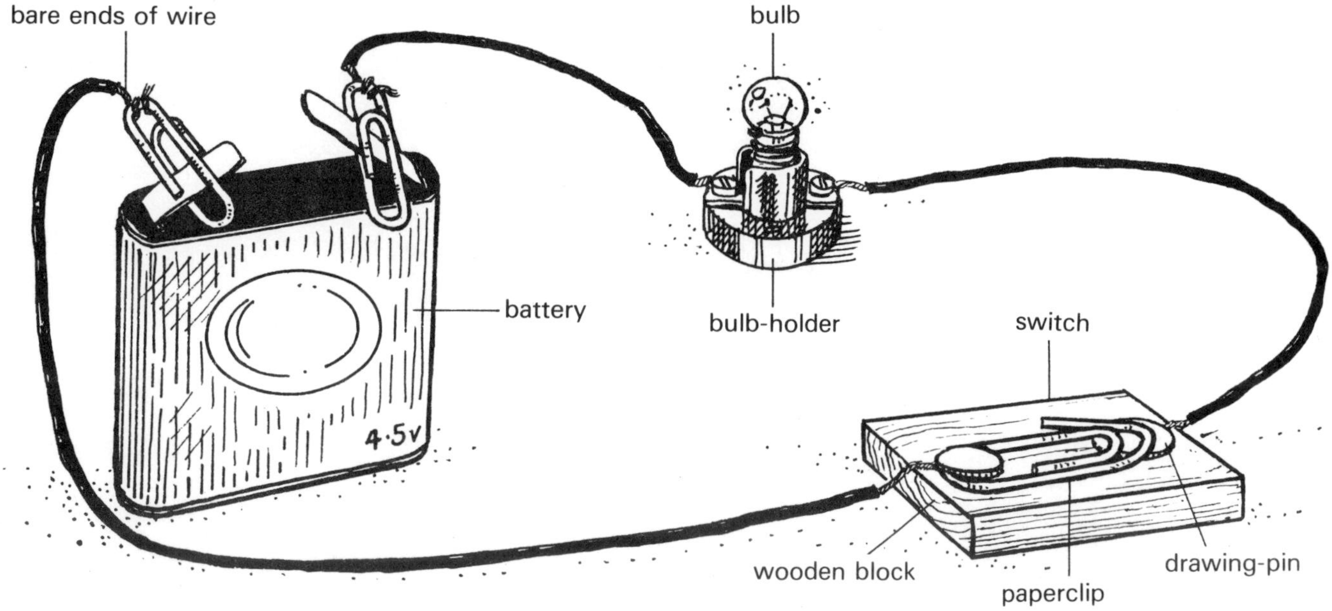

Switch on. What happens?

Switch off. What happens?

Why do you think this happens?

An Edward Arnold Master

28

Start with Science

Electricity:

More circuits

NEVER EXPERIMENT WITH MAINS ELECTRICITY

Name *Class* *Date*

■ **You need:** ☐ battery (4.5 volt) ☐ 2 bulb-holders ☐ 2 bulbs (3.5 volt) ☐ wire (insulated, but ends bared) ☐ switch ☐ paperclips ☐ drawing-pins ☐

You can light two bulbs from one battery.
Make a circuit like this.

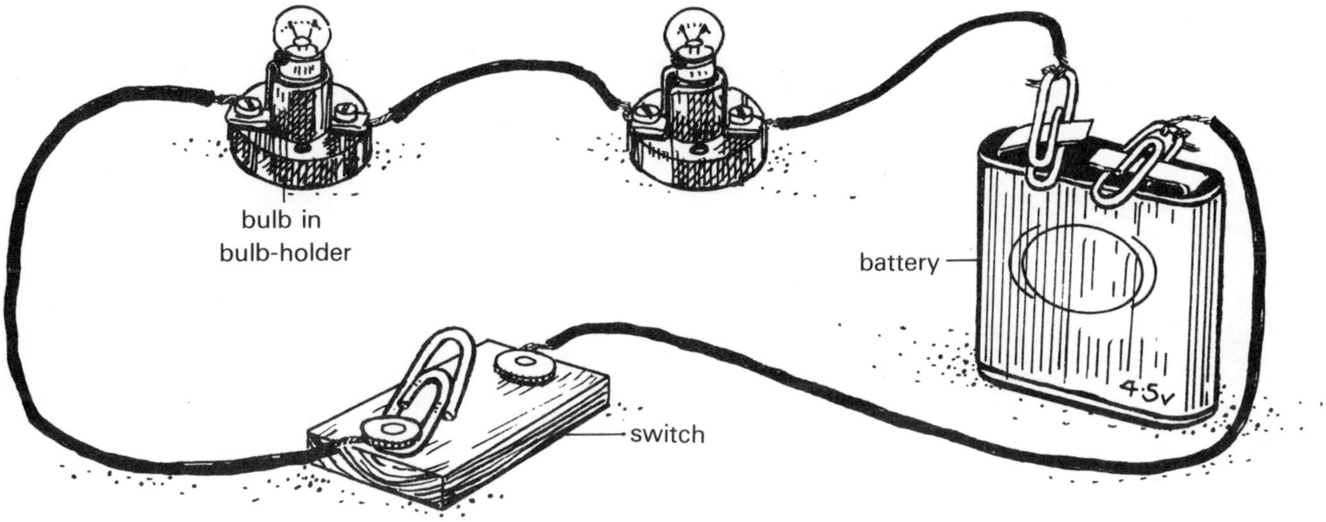

Switch on. What happens? ______________________

____

Unscrew one bulb. What happens? ______________________

____

Connect another bulb into your circuit.
What do you notice about the brightness of the bulbs? ______________________

____

Why do you think this happens? ______________________

____

____

An Edward Arnold Master

29

Start with Science

Electricity:

Morse-code messages

NEVER EXPERIMENT WITH MAINS ELECTRICITY

Name *Class* *Date*

■ **You need:** ☐ battery (4.5 volt) ☐ bulb (3.5 volt) ☐ bulb-holder ☐ switch ☐ wire (insulated, but ends bared) ☐ paperclips ☐ drawing-pins

You can send messages using Morse-code.
It is a mixture of 'dots' and 'dashes'.

A .–	F ..–.	K –.–	P .––.	U ..–	Z ––..
B –...	G ––.	L .–..	Q ––.–	V ...–	
C –.–.	H	M ––	R .–.	W .––	
D –..	I ..	N –.	S ...	X –..–	
E .	J .–––	O –––	T –	Y –.––	

Set up a circuit like this.

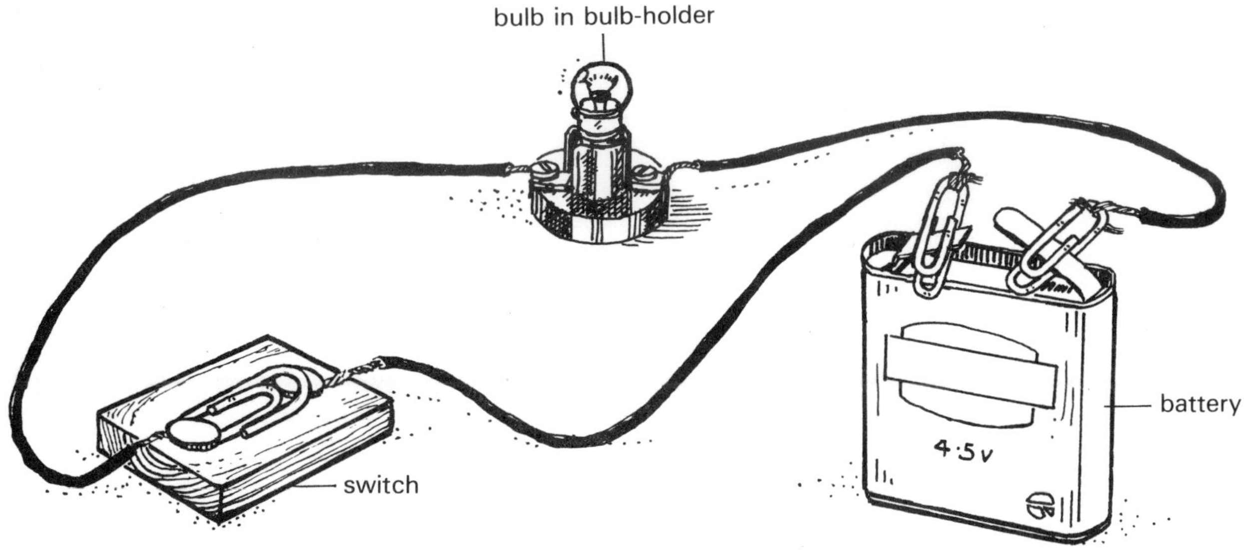

Print a short message here. ____________________

Now write the message in Morse-code. ____________________

Send this message in Morse-code using your signalling circuit.
Use a *long flash* for a 'dash' and a *short flash* for a 'dot'.

An Edward Arnold Master

Start with Science

30 Magnetism:

Using magnets

Name ______ *Class* ______ *Date* ______

■ **You need:** ☐ magnets

Collect some magnets.

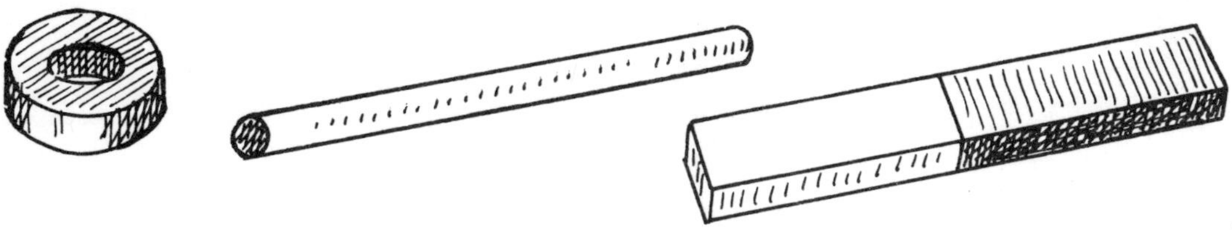

Magnets can pull some objects. We say that magnets **attract** these objects.

Make a list of objects, in your classroom, that are attracted by magnets.
List the objects that are not attracted.

Magnets attract	Magnets do not attract

What do you notice about the 'attracted' list? ______

______

______

______

______

An Edward Arnold Master

31

Start with Science

Magnetism:

A magnetic fishing game

Name *Class* *Date*

■ **You need:** ☐ thin coloured card ☐ scissors ☐ paperclips ☐ plastic bowl ☐ magnet ☐ thin string ☐ short stick

Find some thin coloured card.
Cut out twenty fish shapes.
Fix a paperclip to each fish shape.

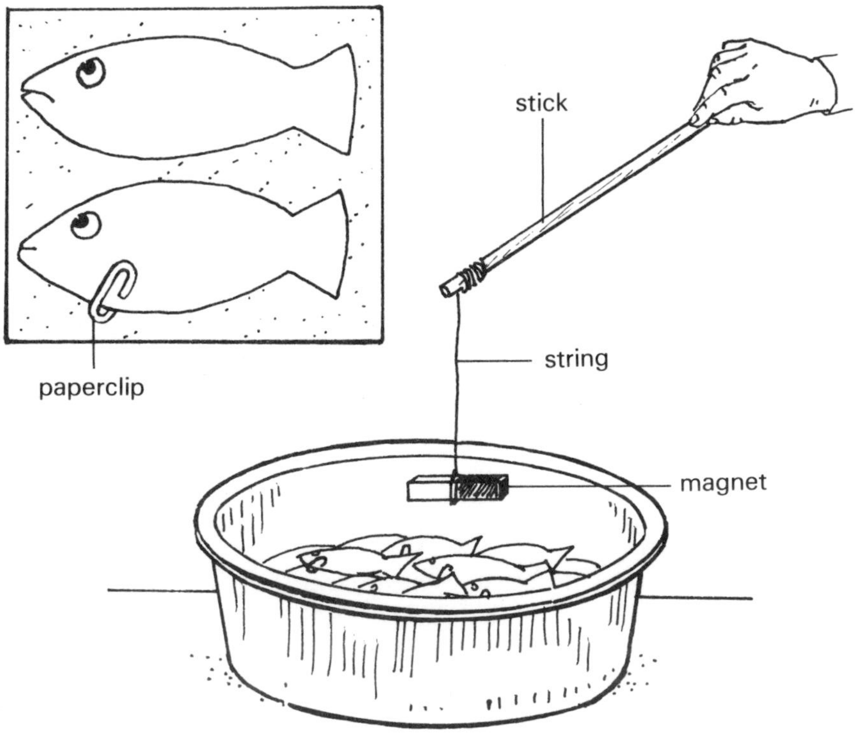

Place all the fishes in a plastic bowl.

Make a fishing-rod using a short stick, some thin string and a small magnet.

Use your fishing-rod magnet to lift the fish out of the bowl

Describe what you have been doing:

____

____

____

____

____

____

An Edward Arnold Master

32

Start with Science

Magnetism:
Make an electromagnet

Name *Class* *Date*

■ **You need:** ☐ wire (insulated) ☐ large iron nail ☐ battery (6 volt) ☐ switch ☐ paperclips

Wind a long piece of wire around a large iron nail. Always wind in the same direction. Connect the ends of the wire to a battery and a switch.

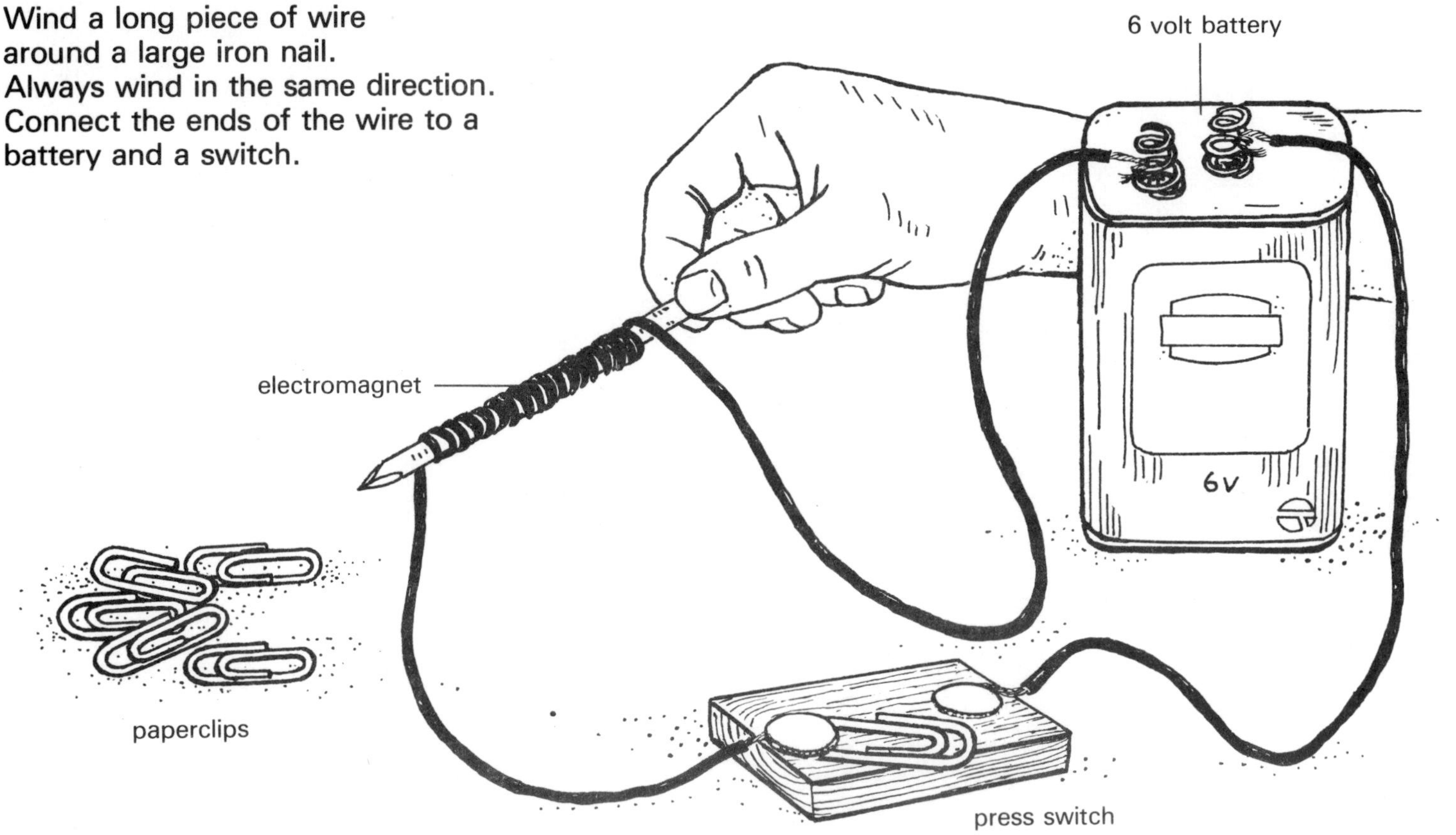

Switch on. Use your electromagnet to pick up some paperclips.
What happens to the paperclips when you switch off?

How could you make your electromagnet stronger?

An Edward Arnold Master

33

Start with Science

Materials:

Strong threads

Name *Class* *Date*

■ **You need:** ☐ threads (various types) ☐ strip of wood ☐ plastic bucket ☐ sand

Collect some threads.
Try to include cotton, nylon, wool and rayon in your collection.
Which is the strongest thread?

Try this test.

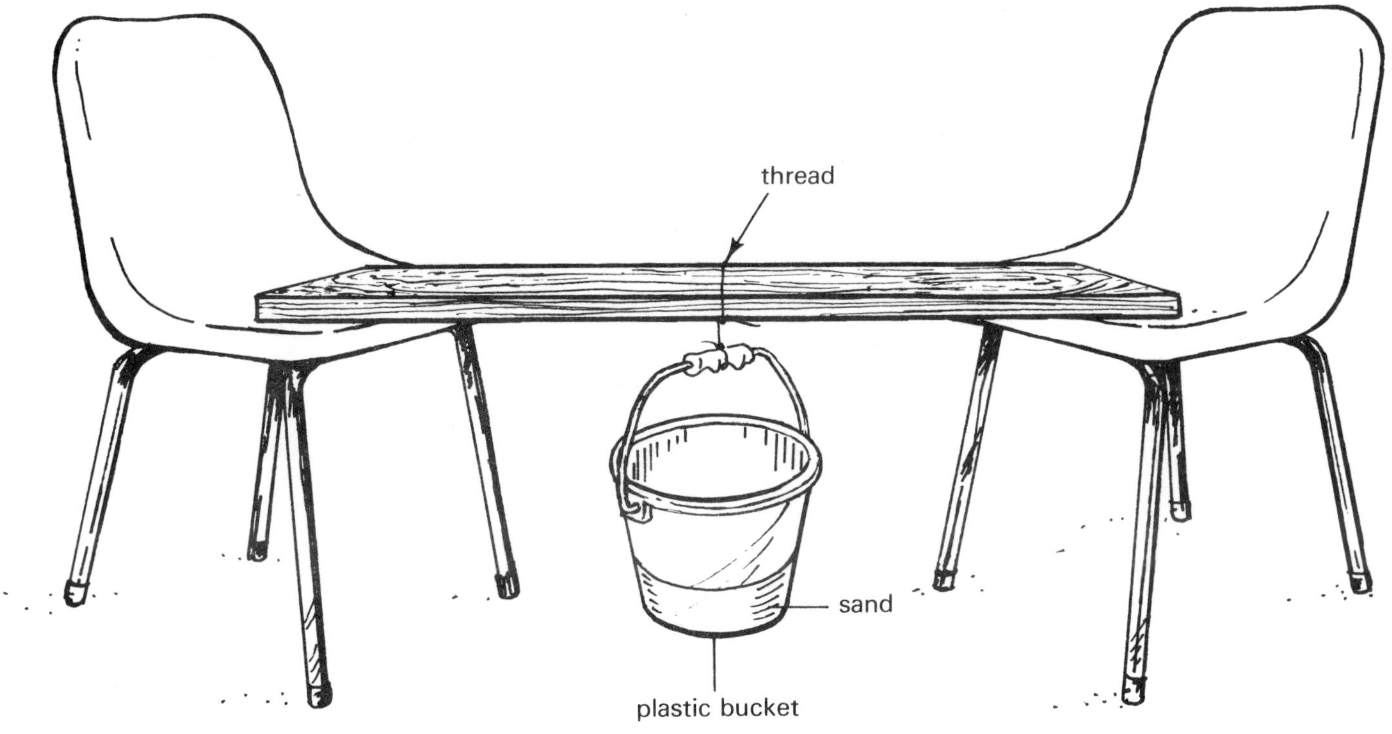

Place a strip of wood across two chairs.
Tie one of the threads to the wood.
Attach the handle of a plastic bucket to the other end of the thread.

Test the strength of each thread by placing sand in the bucket.
Does it always need the same amount of sand to break each type of thread?

How can you test this? __________

Which is the strongest thread? __________

Which is the weakest thread? __________

An Edward Arnold Master

34

Start with Science

Materials:

Make a 'spin-drier'

Name *Class* *Date*

■ **You need:** ☐ washing-up-liquid bottle ☐ scissors ☐ string ☐ wet cloth

Find an empty washing-up-liquid bottle.
Ask your teacher to cut off the top part.
Make many small holes in the bottom section.
Tie a length of string through holes in this section.

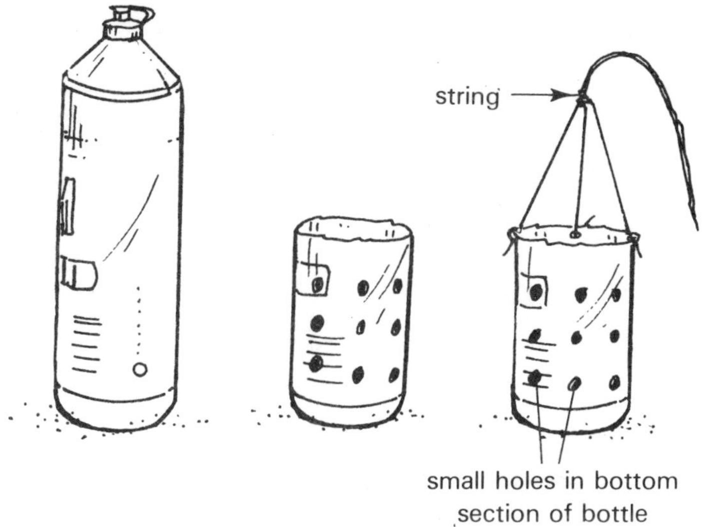

Place a piece of wet cloth in your 'spin-drier'.
Go into the school playground and whirl your spin-drier around.
Notice how the water comes out through the holes

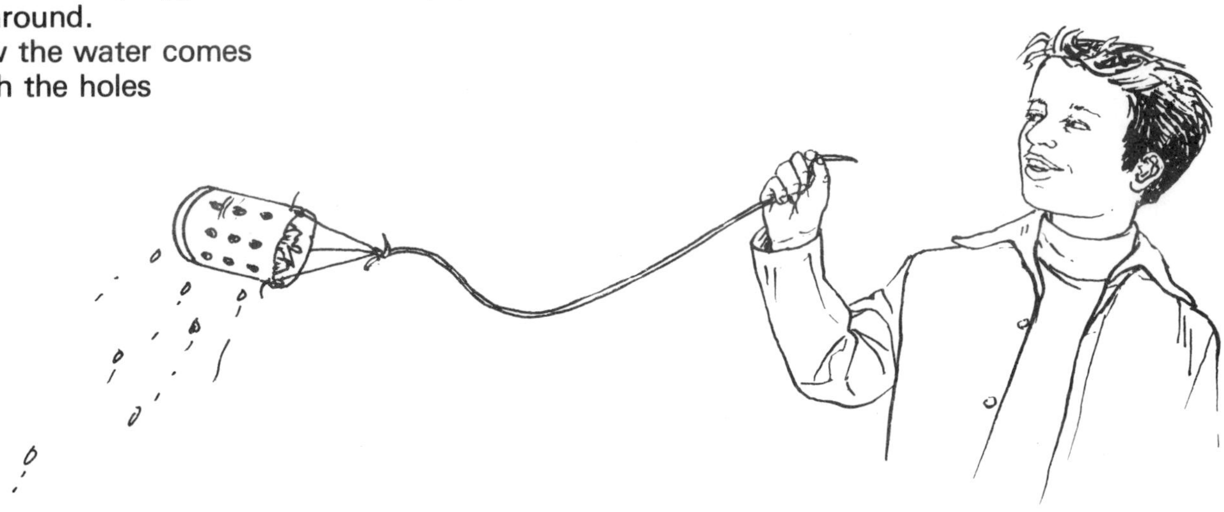

Is this a quicker way of drying materials than hanging them up to dry? How could you test this?

____

____

____

____

An Edward Arnold Master

35

Start with Science

Materials:

Fun with ice-cubes

Name *Class* *Date*

■ **You need:** ☐ ice-cubes ☐ newspapers ☐ clock or watch ☐ cotton wool ☐ polystyrene chips ☐ corrugated cardboard

Try these experiments.

1 Take two ice-cubes.
Wrap one ice-cube in newspaper.
Leave the other cube unwrapped.

Which one will melt first? ________________________________

____

Time them to find out how long each cube takes to melt.

Unwrapped cube melted after ______ minutes.

Wrapped ice-cube melted after ______ minutes.

ice-cube wrapped in newspaper

2 Wrap some ice-cubes in different materials.
Use newspaper, cotton wool, polystyrene chips and corrugated cardboard.
Which material keeps the cubes the longest? Write down your results.

Material	Time
newspaper	
cotton wool	
polystyrene chips	
corrugated cardboard	

An Edward Arnold Master

Start with Science

Small animals:

Looking at earthworms

Name *Class* *Date*

■ **You need:** ☐ earthworms ☐ plastic box ☐ blotting paper ☐ magnifying glass ☐ scissors

Collect some earthworms.
Place them in a small plastic box
with some damp blotting paper on the bottom.
Keep this paper damp while the worms are in the box.
Make air-holes in the lid of the box.

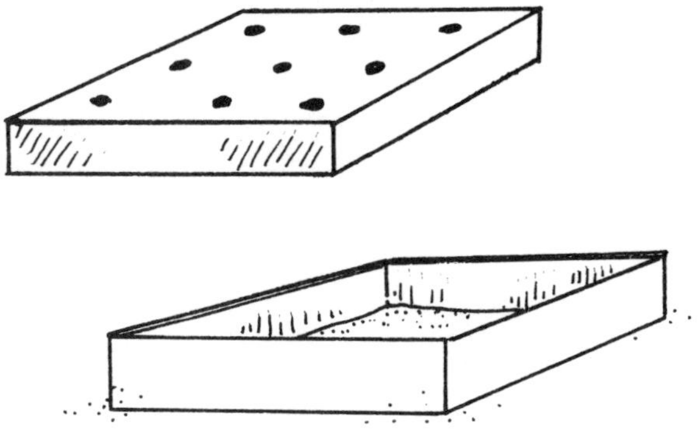

Find a magnifying glass and look carefully at one of the earthworms.

Look at both ends. Are they the same?

What colour is the worm?

What shape is it?

How many segments (parts) does your earthworm have?

Draw your earthworm:

An Edward Arnold Master

37

Start with Science

Small animals:

It's moving!

Name Class Date

■ **You need:** ☐ earthworm ☐ magnifying glass ☐ ruler ☐ paper

Place an earthworm on some paper.
Watch how it moves.
How far does it move in one minute?
How can you find out?

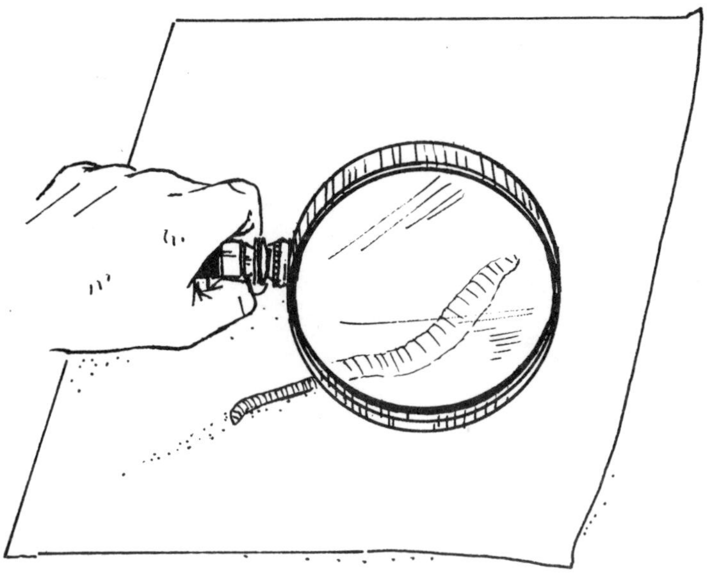

What happens when your earthworm is placed on *damp* paper?
How far does it move in one minute?

Make some measurements:

How long is your worm?

How short can it become?

How long can it become?

An Edward Arnold Master

38

Start with Science

Small animals:

Woodlice (1)

Name *Class* *Date*

■ **You need:** ☐ woodlice ☐ plastic tray ☐ card

You can find woodlice under large stones, old logs or leaves. Try this experiment to find out if woodlice like to live in the light or dark.

Find a plastic tray.
Cover the top of one end of the tray with card.

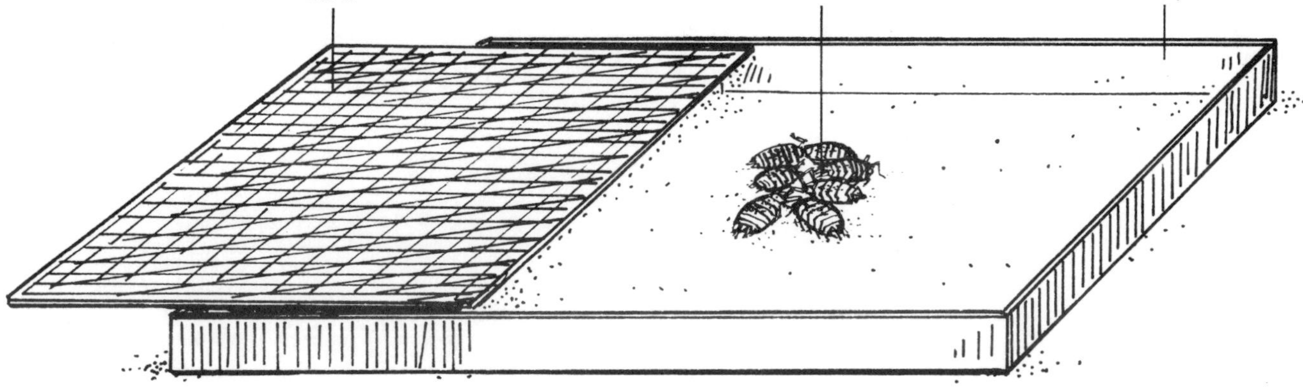

Place some woodlice in the centre of the tray.
Leave them for 10 minutes.

Where have the woodlice moved to? Are they in the light or dark?
Try this experiment a few times. What have you found out?

An Edward Arnold Master

39

Start with Science

Small animals:

Woodlice (2)

Name *Class* *Date*

■ **You need:** ☐ woodlice ☐ plastic tray ☐ cotton wool

Take a plastic tray and place some damp cotton wool near one end.
Place some woodlice in the centre of the tray.
Leave them for 10 minutes.

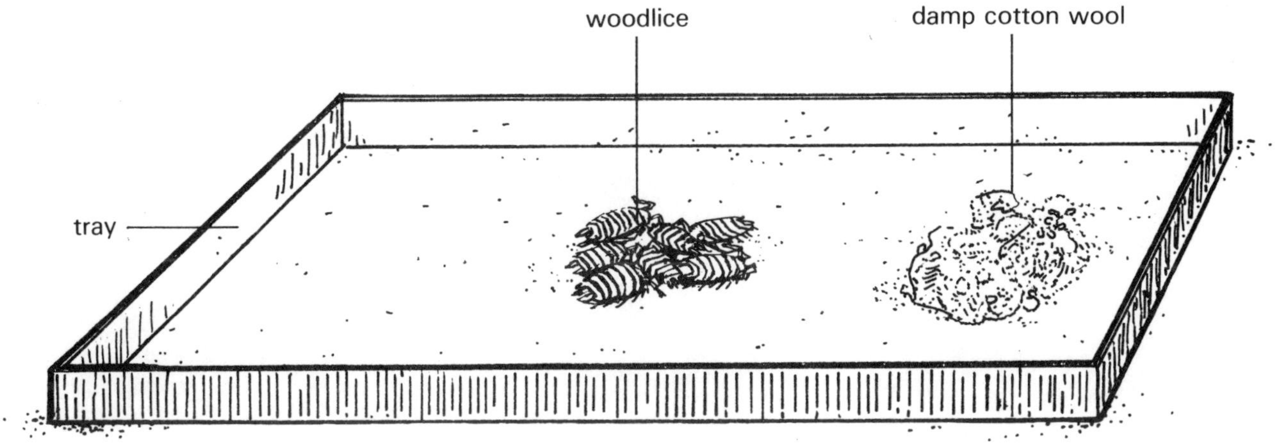

Where have the woodlice moved to?

Try this experiment a few times.
Do the woodlice always move to the same position?
What have you discovered?

How could you find out if woodlice prefer to live in damp, dark places?
Experiment and then describe what you did and what you found out.

An Edward Arnold Master

Start with Science

40 Plants:

Collecting leaves (1)

Name *Class* *Date*

■ **You need:** ☐ leaves ☐ newspapers ☐ card

Make a collection of leaves from different trees.
Press the leaves by placing them between sheets of newspaper.
Leave some heavy books on the newspaper for a few days.

Mount each leaf on a piece of card.
Write the name of the tree beside each leaf.
This leaf chart will help to identify some of the leaves.

Sycamore 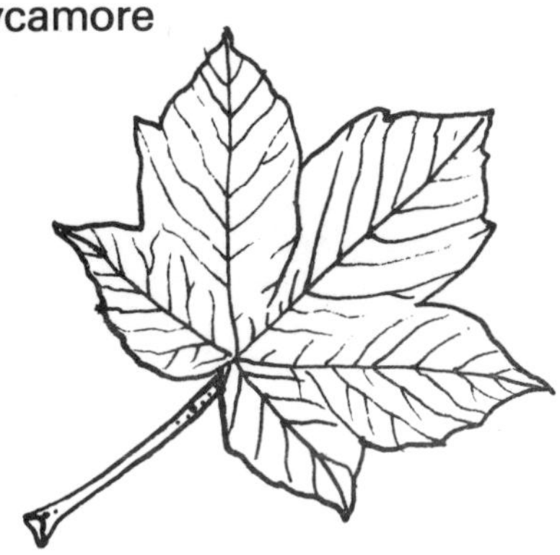	Oak 	Ash 
Horse chestnut 	Hawthorn 	Holly 
Beech 	Plane 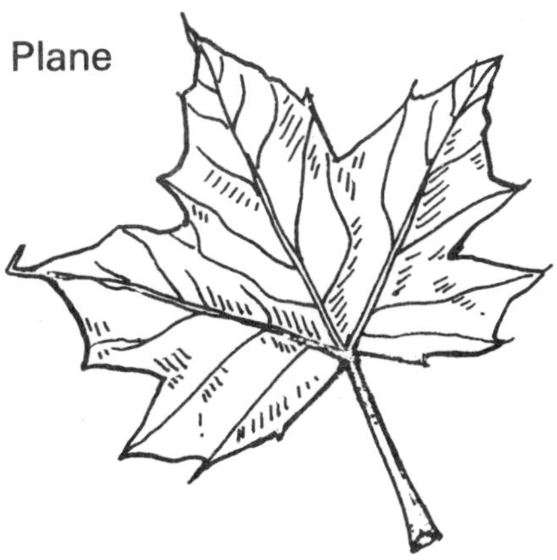	Poplar 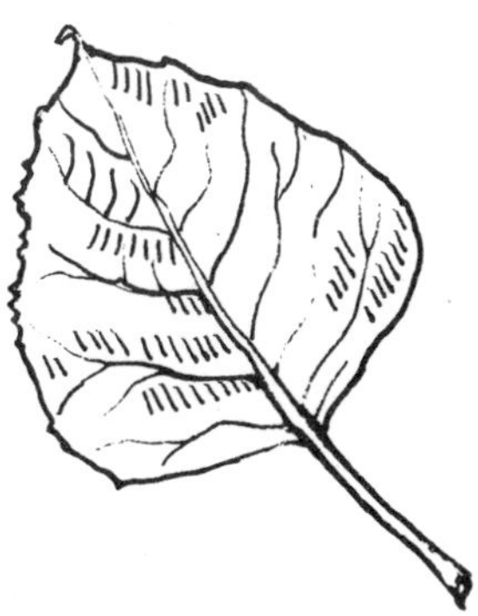

An Edward Arnold Master

41

Start with Science

Plants:

Looking at leaves (2)

Name *Class* *Date*

■ **You need:** ☐ leaf collection ☐ graph paper

Look at your leaf collection.
What colours are the leaves?

What do you notice about
the edges of the leaves?

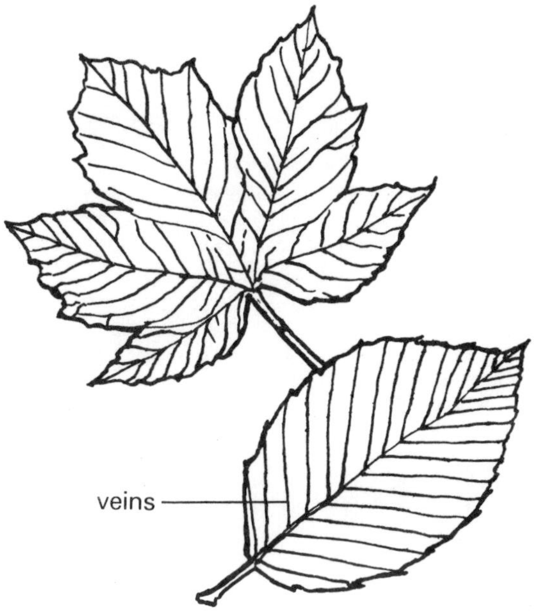

Can you see the veins in the leaves?

Which leaves have the most veins?

Are all the leaves the same thickness?

Which leaf is the smallest?

Which leaf is the largest in your collection?

How could you measure the *area* of one of the leaves?

An Edward Arnold Master

42

Start with Science

Plants:

Tree shapes

Name *Class* *Date*

■ **You need:** ☐ paints and brushes

In winter many trees lose their leaves.
These trees are called **deciduous**.

Look at the *shapes* of some deciduous trees.
Many trees have a typical shape.
Can you see the main branches?
Notice how they grow from the tree-trunk.

Oak

Use some paints and a brush to record your observations.

Try to notice how the trees change during the spring, summer and winter seasons.

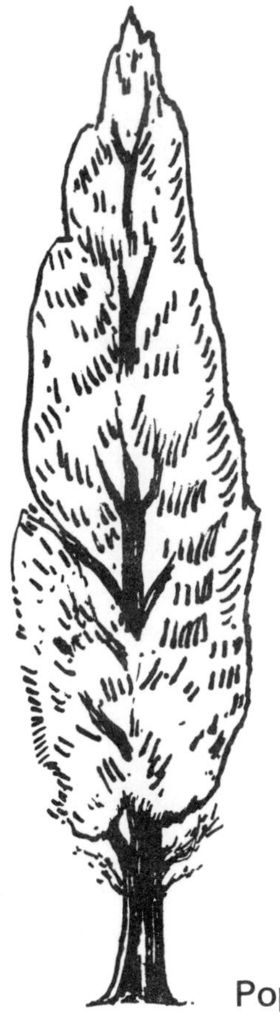

Poplar

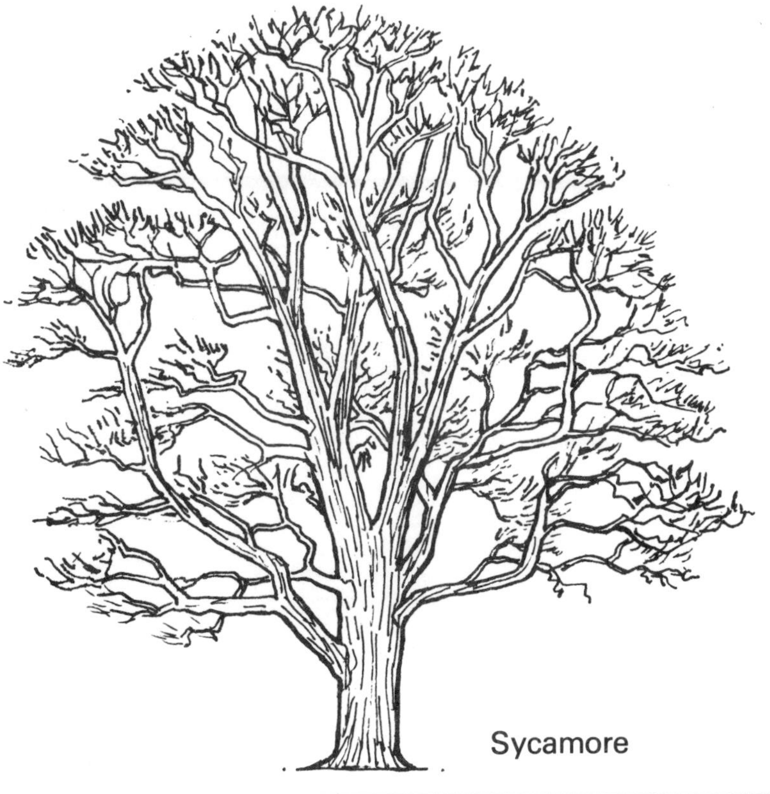

Sycamore

An Edward Arnold Master

43

Start with Science

Plants:

Annual rings

Name *Class* *Date*

■ **You need:** ☐ tree-trunk section ☐ ruler

Your teacher will show you part of a cut tree-trunk.

Look at the bark. How thick is it?

Can you see rings in the wood?
These are called **annual rings**.
A tree grows one ring every year.

How many rings can you count?

How old is the tree-trunk?

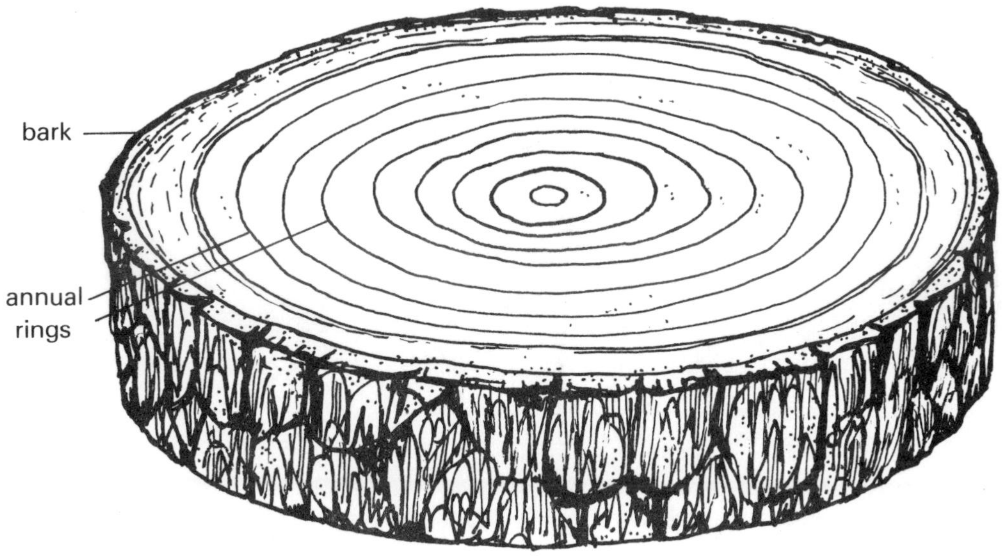

Did the tree grow more some years than others?

How can you tell?

An Edward Arnold Master

44

Start with Science

Plants:

Growing grass seeds

Name ______ *Class* ______ *Date* ______

■ **You need:** ☐ egg-trays ☐ eggshells ☐ cotton wool ☐ grass seeds

Place some empty eggshells in three egg-trays.
Fill each eggshell with some wet cotton wool.
Sprinkle some grass seeds on the cotton wool.

Place one egg-tray near a window in your classroom.
Water these seeds every day if needed.
How many days before the grass starts to grow?

______

Place the second tray next to the first tray, but do not give the seeds any water.

What happens to these seeds? ______

______

______

______

What have you discovered? ______

______

______

______

______

An Edward Arnold Master

45

Start with Science

Plants:

Flowers

Name *Class* *Date*

■ **You need:** ☐ garden flowers ☐ buttercups ☐ jars or vases of water

Collect some garden flowers.
Place them in jars of water and make a display in your classroom.
Put the name of the flower beside each jar.

Write the names of the flowers in your collection:

Find a buttercup flower.
Look for the green **sepals**.

How many are there?

How many yellow **petals** can you see?

There are too many **stamens** to count.
Can you find the **carpels** in the middle of the stamens?

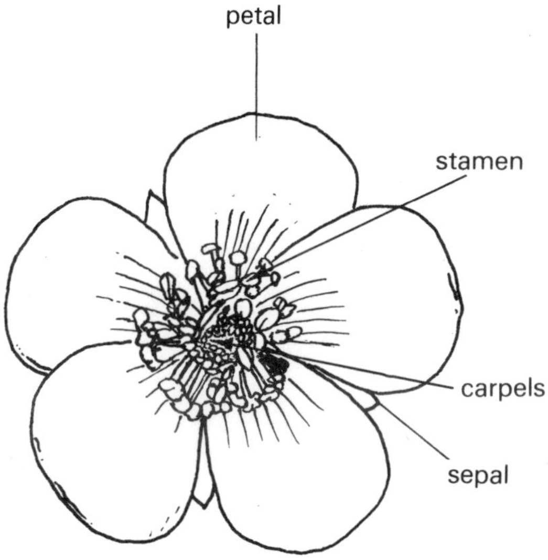

buttercup flower

Look at other flowers. How many petals do they have? What colour are they?

Name	Petals	Colour

An Edward Arnold Master

46

Start with Science

Plants:

Grow an onion in water

Name *Class* *Date*

■ **You need:** ☐ an onion ☐ vase with a narrow neck ☐ water

Find a medium-sized onion.
Find a glass vase with a narrow neck.

Pour water into the vase until it is almost full.
Place the onion on top of the vase.
The bottom part of the onion must be covered by water.
Always keep it like this.

Place the vase and onion in a *cool, dark* place.
How long does it take for **shoots** to grow?

Now move the vase and onion to a *light* place.
Watch how the onion grows.
How much does it grow every week?

Record your measurements:

Week	Shoots measure
1	
2	
3	
4	
5	
6	
7	
8	
9	

Have you made any other observations?

47

Start with Science

Ourselves:

Eyes

Name *Class* *Date*

What colour are your eyes? ______________________

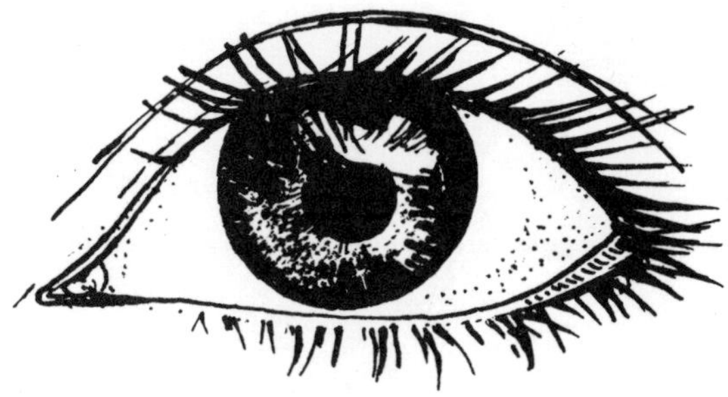

What colour eyes do your friends have?
Use this table to record your results.

Name	Blue eyes	Brown eyes		

Now draw a graph to show eye colours in your class.

Which is the most common eye colour in your class? ______________________

An Edward Arnold Master

48

Start with Science

Ourselves:
Looking at fingerprints

Name *Class* *Date*

■ **You need:** ☐ ink-pad ☐ magnifying glass

Find an ink-pad. *Roll* one of your finger-tips on this pad.
Press your inky finger onto the space below:

my fingerprint

Arch

Loop

Whorl

Composite

Have you made a good print?
Use a magnifying glass to look at it closely.
Look at the other fingerprint patterns.
Which one is like *your* fingerprint?

____

Make prints of your other finger-tips. Which type are they?

An Edward Arnold Master

49	*Start with Science*

Ourselves:
Measure your pulse

Name *Class* *Date*

■ **You need:** ☐ drinking-straw ☐ drawing-pin ☐ clock or watch

Sometimes when you are ill the doctor feels your pulse.
You can measure your own pulse.
Find a drinking-straw and a drawing pin.
Push the drawing-pin through the straw.

Position the drawing-pin on your wrist.
Keep your arm still.
Can you see the straw moving?

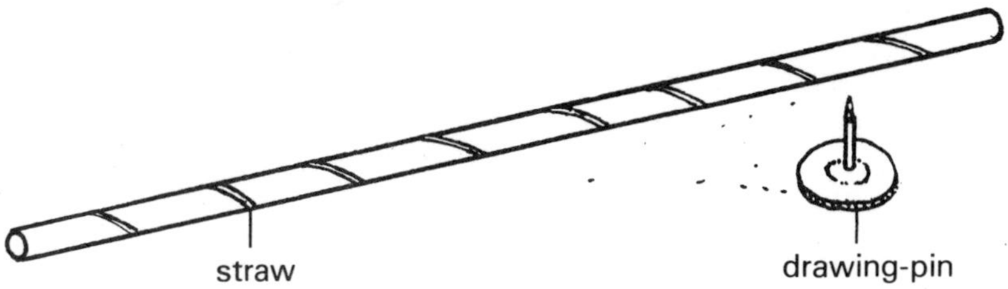

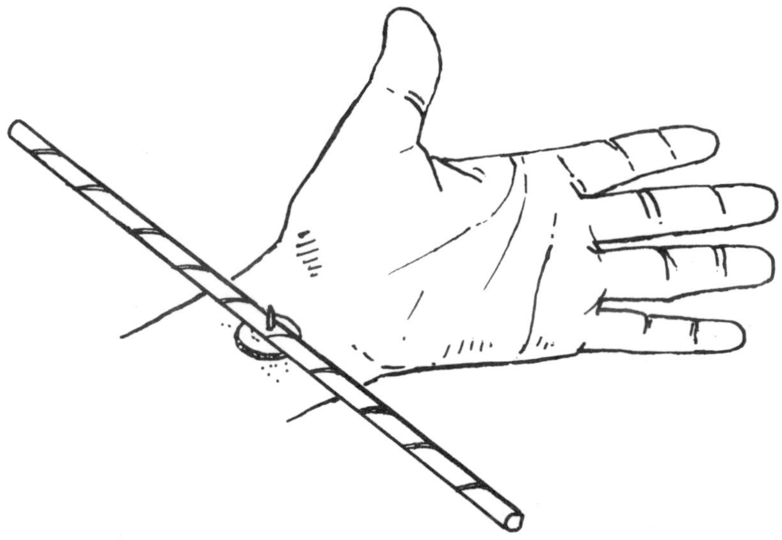

Count how many times your pulse 'beats' in one minute: ____________________

Run around the playground.
Measure your pulse.
How many 'beats' in one minute? ____________________

Finish these sentences:

My pulse was beating f _ _ _ _ _ when I ran around the p _ _ _ _ _ _ _ _ d. This

was because my heart was working harder to pump blood around my b _ _ _ .

An Edward Arnold Master

50

Start with Science

Ourselves:

Test your reactions

Name *Class* *Date*

■ **You need:** ☐ ruler

Work with a partner.
Your partner must sit on a chair.
Hold a ruler so that one end is near your partner's hand.
Let go of the ruler.
Can your partner catch it?

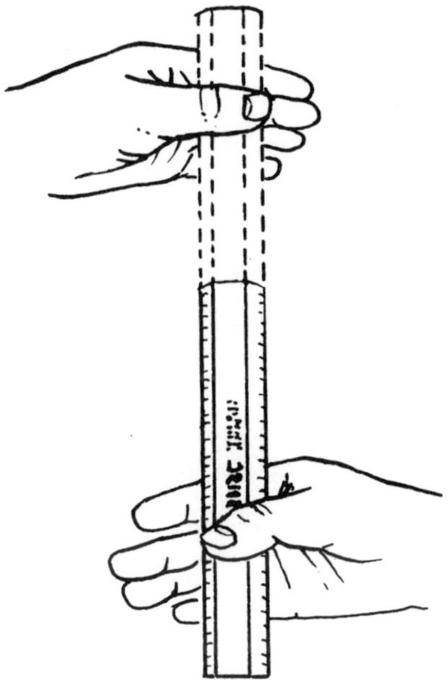

Test your partner three times.
How many centimetres does the ruler drop each time?
Record your results.
Ask your partner to test your reactions.
Record the results.
Test other people in your class.
Look at the results. Who has the fastest reactions?

Name	1st try	2nd try	3rd try

An Edward Arnold Master